Politics of Mining

What They Don't Teach You in School

Edited by Deepak Malhotra

Published by the
Society for Mining, Metallurgy, and Exploration, Inc.

Society for Mining, Metallurgy, and Exploration, Inc. (SME)
8307 Shaffer Parkway
Littleton, Colorado, USA 80127
(303) 973-9550 / (800) 763-3132
www.smenet.org

SME advances the worldwide minerals community through information exchange and professional development. With more than 16,000 members in 50 countries, SME is the world's largest professional association of mineral professionals.

Disclaimer
The papers contained in this volume are published as supplied by individual authors. Any statement or views presented here are those of individual authors and are not necessarily those of the Society for Mining, Metallurgy, and Exploration, Inc. The mention of trade names for commercial products does not imply the approval or endorsement of SME.

ISBN 0-87335-202-5

Library of Congress Cataloging-in-Publication Data has been applied for.

Contents

Preface

An engineer once asked me–"How come we have evaluated more than 50 potential deposits in the last 7 years but have not commercialized one so far?" She was a technically competent engineer but could not understand that socioeconomic and sociopolitical considerations are equally, if not even more, important in commercialization of deposits.

Recent graduates and seasoned professional engineers frequently encounter situations where discussions are held that are counterintuitive to the logically trained engineering mind. "Other" influences, whether they be financial, social, cultural, environmental, or plain old stubbornness by someone higher up in the hierarchy, can sometimes overrule the logical decision-making process. This can be frustrating, to say the least. Therefore, this book explores some of the nontechnical issues that can have a major impact in the mining industry. Some of the topics are unique to the mining industry; others are encountered in other industries. However, all topics are geared to exploring issues that most engineers don't learn in school.

ACKNOWLEDGMENTS

I would like to extend sincere and heartfelt thanks to cochairs Jock McGregor and Leonard Harris, the authors of these papers, and the symposium organizing committee for their valuable contributions to this symposium. Finally, special thanks to Tara Davis at SME and Jyotisna Malhotra of Resource Development, Inc. for their patience, consideration, and valuable input and organization.

Deepak Malhotra

Symposium Organizing Committee

Chair

Deepak Malhotra
Resource Development, Inc.
Wheat Ridge, Colorado

Cochairs

Ronald J. (Jock) McGregor
Vista Gold Corporation
Littleton, Colorado

Leonard Harris
Mining Industry Consultant
Lone Tree, Colorado

Members

William F. Riggs
Cytec Industries, Inc.
Houston, Texas

Bill A. Hancock
Great Western Chemical Company
Murray, Utah

Ed Dowling
Cleveland-Cliffs
Cleveland, Ohio

Han von Michaelis
Randol International, Ltd.
Golden, Colorado

Donald W. Gentry
Polymet Mining Corporation
Houston, Texas

V.I. Lakshman
Process Research Ortech, Inc.
Mississauga, Ontario

SECTION 1

Mining Company Business Issues

Politics of Mining: An Overview

R.J. McGregor[*] and Deepak Malhotra[†]

ABSTRACT

Engineers frequently encounter situations where decisions that are counterintuitive to the logically trained engineering mind are made because of other influences, whether they be financial, social, cultural, or environmental. Such influences can overrule the "logical" decision-making process. The aim of this book is to explore some of the nontechnical issues, which most engineers don't learn about in school, that can have a major impact in the mining industry. This chapter gives a brief overview of this volume and introduces the topics that will be addressed.

INTRODUCTION

This book was originally conceived in 1995 by my coauthor, friend, and colleague Deepak Malhotra. Malhotra has worked in both large and small mining companies, and in recent years, he has been heavily involved with plant audits. In plant auditing, management pays someone to tell them why their employees are not doing the job right. At least, that is often the interpretation of plant personnel, a perception that leads to a situation rife with politics. The chapters that follow were written to enhance our ability to deal constructively with this and other types of day-to-day politics inside and outside the organization, as well as to explore the nontechnical issues that have an impact on our daily working lives. Originally conceived as "political metallurgy," the scope has naturally broadened to encompass other areas in mining and mine development; today, then, this collection of papers is better titled "Politics of Mining." At the end, we hope that this will equip you to better recognize and deal with certain situations as they arise, now and in the future.

Mining is an industry that has many unique factors. For example, we tend to work in remote locations, but the development of a mine in *any* part of the world has major social and environmental implications. As the rights of indigenous peoples around the world are being increasingly recognized and their political strength grows, the need to develop good community relations becomes more important. I do not mean to suggest that international companies in the past have ignored this issue, but now it is a necessity, not an option. In addition, the advent of the Internet has made the problems of mining all over the world more visible; there is no such thing as a "local issue" anymore. Many people in the industry feel

* Vista Gold Corporation

† Resource Development, Inc.

that they're paranoid. But they're not; there really are people out there trying to get you! For a young engineer, first-time exposure to the kind of venom that can come out of public meetings during the permitting process can be quite intimidating.

Mining is capital intensive and high in risk; it requires good test work and development, good studies, and eventually, a leap of faith to spend the money to put a project into production. Every development engineer plays his or her part in that process. Mining is a nonrenewable resource, and in this time when the buzzword is "sustainable development," we have to redefine this phrase to suit our industry.

All the authors have been asked to be candid and forthright in their chapters, for we can learn a good deal from our and other people's mistakes, even though we are reluctant to share experiences that make us look foolish. But I guarantee that you will learn many useful tricks and ideas that you can use for the rest of your careers. And these insights are not taught in a lot of institutes of higher learning. When they start their careers, engineers tend to be specialists with limited exposure outside their areas of expertise. There is a danger in forgetting that what you do interlocks with what others are doing. The sooner you think outside the box and approach your role in its overall context, the more valuable you will be as an engineer and the more likely you will be to get noticed.

When first joining a public gold mining company, a junior mining engineer was asked what the company's goals were–"to mine and produce gold" was his quick answer, but it was wrong. The mining and production of gold was not the goal, it was the means of achieving the company's goal, which is to make money for the shareholders. We hope that this book will make you think about the bigger picture and think outside the box.

MINING COMPANY BUSINESS ISSUES

Our first section, of which this paper is part, explores mining company business issues. As I mentioned in the introduction, a mining company's role is to make money. We do this by mining. We then have to decide what to mine. Some companies are single metal companies, such as Barrick, for example, or my own company, Vista Gold. Others mine polymetallics and maybe some industrial minerals. Some companies change over the years, because it is important for management to constantly reevaluate the company's business plan. Some companies look to vertical integration. For instance, Anglo Gold recently bought a large jewelry manufacturer. Other decisions may involve in which and in how many countries to operate. Small companies can get into trouble by trying to manage projects in too many foreign countries, particularly if there are many cultural differences.

Having decided on a strategic business plan, a business structure to execute your strategic plan must be developed. Many properties are owned as joint ventures, sometimes run by management committees. Often, royalties are paid, but there are many different structures of mine ownership. These can have a significant political impact on the mine operation. There are also decisions to be made on how to operate the mine. Will you use contractors or in-house personnel? Will you train locals or import expatriates? All these decisions must be made during the development phase. In addition, it is important to decide what return on investment the company will accept for a project. It will be argued that mining companies accept far too low of a return for the risk that they take. Finally, on old or new projects, it is necessary to do due diligence to research and verify many different issues. Several consulting companies perform technical due diligence; however, nontechnical due diligence can have as big if not a bigger impact as the technical issues. All are ignored at your peril.

COMPANY POLITICS

Next, we have two sections on company politics, which focus on areas that people don't think a lot about, but that can have a major impact on their lives during the trials, tribulations, and growth of the company. Each company develops a culture of its own, and if you don't fit into that culture, your working life can be unpleasant. When companies merge, there is often a clash of cultures, which has a major effect on the people, their ability to perform, and their job satisfaction. Most companies aim to grow, but growth can bring its own problems and size can often dictate the management philosophy. Sometimes functions are centralized at the head office. Sometimes each mine manages its own affairs. Often, there is a neverending cycle of change, depending on the management at the time. All these issues can dramatically affect how an engineer does his or her job. Another political issue that is very important to address is interdepartmental rivalry, which often occurs in operating mines. We will discuss how to prevent it from being destructive and make it constructive.

The role of the engineer in the organization is another very important topic; the engineer works with a lot of diverse disciplines, such as attornies, accountants, and environmentalists, among others. It is important that the engineer understands his or her fundamental role in the organization. Finally, we have some case histories of things that have gone wrong.

When we turn to intercompany politics, we look at communications and their importance. For example, how do different companies deal with project development? Do operators have the skills to build a project or do you need a separate development group? Companies have different ideas on these issues. What is the role of consultants and technical auditors, and why does management always seem to think that they're smarter than their own people? How do engineers develop into managers? Thinking outside the box certainly helps, but engineers that are specialists can often get "pigeonholed." How do you avoid this and get the opportunity to develop?

SOCIAL, COMMUNITY, AND CULTURAL ISSUES

Section Three covers a wide array of topics under the title "Social, Community, and Cultural Issues." Because mining is an international industry and many of you have traveled or will travel overseas to work, it is very important to understand some of the challenges involved. Local community support and involvement is no longer an option, and we'll be reviewing how to develop good community support for projects. Indigenous people in all countries are developing strong political voices, but are often outside the mainstream of the sociopolitical structures of their own countries. There are many examples of mines being shut down by civil unrest and civil disobedience regardless of the technical merits of the mine. So social involvement and community support is essential in the development of any mining operation. In most developing countries, many organizations are at work managing hundreds of millions of dollars in aid programs. The majority of these organizations are called nongovernmental organizations and they are a force to be reckoned with both here and abroad. Some firsthand experiences of the good, the bad, and the ugly will be discussed in the paper entitled "Nongovernmental Organizations: Friend or Foe?"

The Internet is an amazingly powerful tool for disseminating unedited information worldwide at very little cost in the blink of an eye. Unfortunately, in the wrong hands, this same immense Internet can be a powerful anti-mining tool; this issue will be discussed. Finally, transient labor problems at home and abroad are a big issue. At the time of construction of a new mine, it's important to develop a good relationship with the local communities, but at the same time there is a huge influx of transient labor whose job is to build the mine and leave; they have no long-term stake in the community or the project. This can lead to some massive public relations problems, which need to be addressed.

TRAINING AND TECHNOLOGY TRANSFER ISSUES

The final section covers training and technology transfer issues, both at home and abroad. Operator training is essential to achieve production goals on a new project, but it's an area that's often overlooked. Technology transfer as technology advances is an important issue, particularly overseas. One issue that is getting more and more attention recently is mentoring, an important career development tool for big and small companies. Another very important subject that will be discussed is the use of orientation programs, which should include both workers and their families for overseas employment. The pros and cons of expatriate versus local management will also be discussed.

CONCLUSION

A final point should be made—we all have something to contribute to the discussion here; we all have anecdotes and experiences that our colleagues can benefit from. Don't be shy in sharing them with us. They may be embarrassing, but we do learn from our mistakes, and others can learn from those mistakes as well.

Due Diligence: Technical and Nontechnical Considerations

Peter Keppler* and Stephen D. Bundy†

ABSTRACT

"Due diligence" is a review or audit of a facility or business to assess and quantify risks and operating status before an acquisition. Due diligence findings can be used to negotiate the price, general terms, and indemnification provisions in the purchase and sales agreement. In addition to assessing outstanding liabilities, necessary capital improvements, and the adequacy of operating budgets, due diligence should include "nontechnical" considerations, such as local community acceptance and relations, and the effect on the company's public image and operations. This paper will discuss how these considerations can be addressed in the due diligence process and factored into the transaction. Also, innovative means for risk reduction and risk transfer, during and after the transaction, will be discussed.

INTRODUCTION

Due diligence is performed on business and facility acquisitions to verify sellers' claims, usually before finalizing the purchase and sales agreement. Traditionally, due diligence has focused on aspects considered to most directly affect the purchase price, such as the raw data underlying the current financial statements and the evaluation of real estate, equipment, inventories, and reserves. Current due diligence practices have been expanded to include a number of the many other aspects inherent in the growing complexity of modern business. Because of the potentially significant costs now associated with environmental matters, environmental due diligence has become an integral part of most companies' due diligence efforts and is almost universally demanded by lenders and investors.

Environmental due diligence focuses on identifying and quantifying financial liabilities that result from existing, pending, or potential regulatory or legal actions, along with additional capital or operating expenses, or both, necessary to comply with existing or pending regulatory requirements. Environmental due diligence, then, can be defined as a "review or audit of a facility or a business unit that assesses and quantifies risks and operating status prior to an acquisition." Given today's regulatory and litigious atmosphere, environmental due diligence is a major factor for potential buyers of mines and mineral processing facilities.

* Attorney at Law

† President, EnviroGroup Limited

Environmental due diligence encompasses both technical and nontechnical aspects and, in today's society and business atmosphere, both are equally important. This paper will discuss both aspects and introduce the concept of risk transfer as an option to facilitate handling environmental risks and increase the likelihood of a successful transaction.

TECHNICAL CONSIDERATIONS

Environmental assessments of industrial properties have become standardized (e.g., see American Society for Testing and Materials [ASTM] E1527-97 [Phase 1]; ASTM E1903-97 [Phase 2]; and ASTM E1528 [Transaction Screen Process]). These ASTM standards represent a good starting point for environmental due diligence, but because of the potential costs of environmental impacts, some industrial acquisitions, including and perhaps especially mining operations, should be subject to a higher level of review.

The term "Phase 1" refers to the initial stage of due diligence conducted by reviewing records, interviews, and a visual inspection of the property and facilities. A Phase 1 due diligence of a mining operation should, at minimum, include:

- A database search for all regulatory actions concerning the facility and vicinity
- An inspection of all environment-related facility records
- A physical facility inspection
- Interviews with key management and employees involved with environmental aspects
- Phone interviews or meetings with representatives of governing administrative agencies
- An analysis of the facility's status under all applicable regulatory programs
- A review of all existing, pending, or potential regulatory actions
- An examination of all existing, pending, or potential environment-related litigation
- An evaluation of potential reclamation and remediation costs
- A cost estimate of potential liabilities
- A cost estimate of potential capital costs and increased operating costs

In the case of a stock or business purchase (versus an asset purchase), the liability portions of the analysis should be expanded to include, and additional database searches should be conducted for, all former operating properties, all named entities under which the company conducts business (or has conducted business in recent years), and all existing and recent joint ventures.

Even though the tasks involved in a "typical" Phase 1 can be outlined in a general sense, it is an investigation, and consequently, once initiated, it will become dynamic. Some lines of questioning are either quickly satisfied, or terminated, either because of obvious lack of significance or lack of information. Other lines lead to further questions and may become time-consuming. A thorough Phase 1 investigation of a major mining operation can easily cost more than $15,000.

The term "Phase 2" refers to a follow-up investigation consisting of physical measurements and environmental media sampling designed to answer questions raised in Phase 1. Although general guidelines exist for Phase 2 investigations, each should be custom designed to be specific to the findings of the Phase 1 investigation. Consequently, no cost, or range of costs, can be realistically estimated until Phase 1 is completed. Theoretically, the decision on whether to conduct a Phase 2 investigation should be based on the likelihood of obtaining meaningful information for a cost that is advantageous in relation to the potential risk being investigated. In practice, transaction deadlines often dictate whether, and to what degree, follow-up investigations are possible.

NONTECHNICAL CONSIDERATIONS

The nontechnical considerations in due diligence for mining operations can be characterized as follows:

- Local community acceptance or attitude toward the operation or facility
- The public image of the operator
- The relationship of the operator with various stakeholders, including regulatory agencies, the local community, political leaders, employees, and lenders
- Management's concern about environmental compliance and health and safety
- The affect of public perception of the mine operator on stock valuation

The following discusses each of these considerations and presents suggestions on how they can be addressed in the due diligence process.

It is generally accepted that local community support is needed in order to permit a new mining operation or major industrial facility. Failure to obtain such support will usually stop even a sound technical proposal that has little or no long-term adverse environmental impacts. We have all heard or read about "horror stories" where a major development with substantial economic benefits has been delayed or killed by local residents or "public interest" groups opposed to resource development. Often, environmental groups that have the resources for launching a major public relations campaign support such opposition. With access to the Internet, opponents to the project can disseminate information very quickly and generate what appears to be widespread opposition to the proposal.

For an existing mining operation, comprehensive due diligence will assess the local community acceptance and support for the mine. This can be done by talking with the local newspaper editor, political leaders (such as the mayor or county commissioner), and other community leaders. Answers to a few key questions can give a good indication of community attitude toward the mining operation. Such information is invaluable to the buyer in determining the future life of the mine, the feasibility of expansion, and the cost of final closure. If the local community is not supportive of the mining operation, it will be difficult and costly (i.e., time-consuming) to expand the operation or even maintain the status quo, and final reclamation will have to meet high public expectations.

The mine operator can foster community support in various ways, including keeping communications open, supporting local services and charities, and holding a periodic open house where the local residents are welcome to tour the mining operation and learn firsthand what goes on behind the fence. If the buyer expects to carry out such activities to maintain a viable operation, the cost of these public relations efforts should be included in determining the purchase price.

Thorough due diligence will assess the public image of the mining operation and place a value on such public perception. This can also be determined through interviews with community leaders, regulatory agencies, and other public representatives. The benefit of a positive public image is hard to quantify, but it clearly adds to the value of the business or assets being acquired. Likewise, a negative public image reduces the value, or increases operating costs as a result of closer public scrutiny and government agency oversight, and possibly because of more stringent regulatory (permit) requirements as well.

Assessing the relationship of the mine operator and management with stakeholders should be incorporated into a due diligence review. Major stakeholders include employees, stockholders, lenders, and suppliers and customers. All these parties have a considerable stake in the business and should be consulted for their perspective on the viability of the operations and practices of the management team. Putting a value on good stakeholder relations and business practices is difficult and requires thoughtful evaluation, but it can be done by consultants experienced in business valuation and due diligence of mining companies and mineral properties.

Assessing management's attitude and concern about environmental compliance and health and safety is also part of comprehensive due diligence. Lack of an environmental management system and health and safety program indicates bad management and poor business practices. A good environmental, health, and safety (EH&S) system that meets generally accepted industry standards adds considerable value to the business; lack of such a system will discount the value.

Finally, it is important to consider the impact of the company's public image on stock value. How the business is perceived with respect to environmental performance and community relations can have a major effect on the market value of the company's stock. A company with a good reputation for minimizing the environmental impact of its operations and working with local communities will generally have a higher market valuation than a company with a negative public perception of its environmental performance. As we are all painfully aware, the mining industry generally still has a poor public image for practices that largely belong to the past, but still tend to depress the value of mining companies' shares.

QUANTIFYING NONTECHNICAL CONSIDERATIONS

It is difficult to assess the nontechnical aspects of a due diligence review and put a dollar value on these aspects. How much is a good public image worth? What is the value of a good relationship with government agencies and political leaders? What is the value of an environmental management system? Like determining the value of "good will" for an ongoing business, a qualified due diligence team or consultant can establish a value for these intangibles for a mining company or for individual operations.

A favorable image and community support is critical today for new mining projects, as well as for existing mines and processing facilities, particularly if operational expansion is planned. The added value of good community and other stakeholder relations is hard to quantify, but a poor public image will usually result in a substantial discount of the purchase price of the business. Because of the negative impact on the operation and the resulting increased operating and maintenance costs, bad public image and relations can discount the purchase price by as much as 5% to 10%. A positive image and stakeholder relations may increase the purchase price or stock value by several percentage points.

The price of a mine or mineral processing facility is usually calculated by determining the present value of the proven mineral reserve in the ground less recovery costs, or the added value of the milling, smelting, or refining process. Any additional operating and maintenance costs for mineral extraction, processing, and reclamation will deduct from the purchase price. If the mine operator has not established and maintained good community relations, it is likely that some additional operating costs are being incurred. For example, the mine manager and staff may have to spend time responding to public concerns and questions or defending company practices that otherwise would not have to be addressed. In addition, the mine may be required to meet pollution standards more stringent than those governing other operations because of government and public concern about lax environmental controls or poor management practices.

RISK TRANSFER—MANAGING RISKS IN MERGERS AND ACQUISITIONS

A relatively new concept in mergers and acquisitions is transferring environmental risks using insurance instruments. The discovery of environmental risks can become a large hurdle in the transaction. It would appear logical that the sellers retain all risks and liabilities that are a result of operations under their ownership, but the situation is rarely that straightforward. The risks identified are typically a mixture of known, probable, and possible risks. Further, because most environmental due diligence investigations are conducted under restrictive deadlines, there is the distinct possibility that significant, but totally unidentified risks, exist. Therefore, in all but the simplest cases, a great deal of uncertainty will surround the risks themselves.

Compounding these uncertainties is the inherent difficulty of estimating the potential costs to mitigate the risks. Typically, even the known risks are not characterized sufficiently to allow accurate cost estimates, especially given the fact that resolving most environmental situations will require satisfying one or more government agencies.

Traditional means of dealing with these uncertainties, such as price reduction, escrow accounts, and total or partial liability indemnifications, among others, are often inadequate to handle these complex situations. Whereas buyers are understandably reluctant to assume risks created by the seller, the seller desires a reasonable return for the assets and cannot agree to indemnify the buyer against all past, present, and future identified and unidentified risks. Consequently, negotiations concerning this aspect are predictably difficult and many are unsuccessful, resulting in either terminating the transaction, or forcing one of the parties into an unrealistic and unsound agreement.

Recently, a number of insurance underwriters have developed instruments that can be used to handle some of these uncertainties and streamline the transaction. These instruments have various names, depending on the company that offers them; however, they fit into two basic categories. The first is designed to cover known risks. These are the so called "cap" policies and are designed to cover expenses over and above current estimates. The second is designed to cover unknown risks.

Both types of instruments are custom tailored to each transaction. Further, this area of the insurance industry is actively evolving. Therefore, the assistance of a knowledgeable insurance broker will be required to determine the probability of designing an instrument that will facilitate any given transaction.

In the right situation, a properly designed instrument or combination of instruments can offer alternatives whereby the buyer, seller, or both can transfer a definite portion of the potential risk to insurance carriers for a known fee, reducing uncertainties and rendering an otherwise untenable situation to an acceptable business risk for both parties.

CONCLUSION

The technical aspects of due diligence investigations are generally well defined. Several good references or guides are available. The nontechnical considerations and risks are more difficult to address and quantify. Estimating the value of good or bad community relations and public perception of the company or operation, and how these factors affect the purchase (stock) price, can be a daunting task. Such matters should be assigned to experts knowledgeable in business valuations and "good will." In addition, new insurance instruments have recently become available for transferring known and unknown risks in mergers and acquisitions. These instruments can provide assurances to the seller and buyer and limit the liabilities of the parties, thereby enabling the transaction to go forward.

CHECKLIST OF NONTECHNICAL CONSIDERATIONS IN DUE DILIGENCE INVESTIGATIONS OF MINING OPERATIONS

✓ Assess local community support and acceptance.

- Interview community leaders, including newspaper editor, mayor, county commissioners, and others.

✓ Assess public image and perception of mine operator.

- Interview community leaders.
- Talk with regulatory agencies.
- Talk with other public representatives, including environmental activists.

✓ Assess relationship of mine operator with stakeholders.
  - Interview employees, suppliers, customers, lenders, and stockholders.
✓ Determine management view and approach to EH&S programs.
  - Talk with employee representatives, regulatory agencies, and environmental groups.
  - Determine extent of regulatory agency oversight and concern; frequency of inspections.
  - Review environmental management system; compare to industry standard (International Standards Organization [ISO] 14000).
  - Review safety records (lost-time accidents).
✓ Assess impact of public image on stock value.
  - Add value for good public image and community support.
  - Discount value for poor public image and bad press.
  - Quantify value, similar to business good will.
✓ Evaluate rsisk transfer through insurance instruments.
  - Environmental liability insurance.
    - Cap insurance
    - Unknown risk/liability insurance
  - Other business risk insurance.

Gold Medal Performance

William Mark Hart*

ABSTRACT

Newmont Mining's executive management launched an initiative in 1999 that was designed to engage the entire workforce in a continuous effort to systematically drive inefficiency out of the corporation. Built on a foundation of teams and teamwork, Gold Medal Performance (GMP) has had a significant and positive impact on the reduction of costs. At the same time, it has increased productivity, improved cash flow, and helped Newmont to further enhance health and safety programs.

Currently, more than a hundred teams are working on improving line-balancing production and external communications at all levels of the corporation. The training and rigorous set of tools available to and used by these teams were designed to help them improve communications, identify inefficiencies and their financial impact, develop and implement solutions, and finally, to measure the results of the process enhancement. GMP harnesses the energy, creativity, and productivity of the entire workforce.

BACKGROUND

In 1921, William Boyce Thompson founded Newmont as a holding company to invest in worldwide mineral, oil, and related companies. He chose the name Newmont as a contraction of New York and Montana because, as one biographer puts it, "he grew up in the latter and made his money in the former." From the beginning, Newmont focused on creating shareholder value.

Today Newmont faces a challenging business environment fraught with low gold prices and increased regulatory pressures, along with social and political issues both at home and abroad. Newmont successfully adapted itself to changing and adverse conditions in the past and realized value by seizing new opportunities. With GMP, Newmont continues this trend in today's business environment.

INTRODUCTION

Newmont, like many other companies, has tried several process management initiatives during the last decade. Each one added some value and helped to develop a culture that is more receptive to Newmont's GMP initiatives. Newmont's executive management has been

* Newmont Mining Corporation

unwavering in its support of the change concept, much of which is based on General Electric's (GE) principles for quality change management.

What sets GMP apart from past initiatives is that everyone is involved and is truly considered an important contributor. This allows for all creative ideas to be taken into consideration. GMP is all about breaking down bureaucracy and rigidity contained in today's business environment.

GMP has a major difference from many other process management methods. A GMP team consists of people who are the implementers of the initiative. This concept allows for direct work experience input and gets a true buy-in from those who are actually going to do the job. The objective is to involve each person in the change process.

GOLD MEDAL PERFORMANCE

GOAL: For all employees to systematically drive inefficiency out of the entire company.

To achieve the goal, a set of tools has been developed and deployed at all levels of the corporation. These tools focus on a disciplined and systematic approach to teamwork, communications, and measurements. They were developed by a number of Newmont global teams that also adopted or adapted some proven techniques and methodologies used and benchmarked by other companies.

Many of these tools simply apply common sense. They build on what is already being done in an ad hoc fashion, but add discipline to the process. The best practices are shared globally for the benefit of the corporation. The value is created when these tools are applied consistently and in concert at all levels of the organization. Each tool utilizes a structured and disciplined process that is documented in PowerPoint for easy presentation to the workforce. Some elements of the tools utilized, some of which are described later in detail, include:

- Newmont's vision and values
- Change acceleration process
- Process mapping
- Dollarization
- Critical performance indicators
- Line balancing
- Risk management
- Rewards and recognition
- Teams and teamwork
- Communication sessions
- 3Ws (who is responsible for doing what by when) and AOCs (activity schedules)
- Value-creation opportunities
- Measurements
- Standard operating procedures
- Cost-out goals
- Workouts

NEWMONT'S VISION AND VALUES

Newmont has provided unwavering support for GMP from the start, from the grassroots to the executive level. A sampling of approximately 300 people from across the company helped to develop a vision and enhance Newmont's values. Newmont's vision is to be the world's most valued mineral resource company. The remainder of the vision statement, below, shows *how* that vision is to be realized.

Newmont's Vision Statement

Newmont will be the world's most valued mineral resource company, creating shareholder wealth through aggressive exploration, technological innovation, operating excellence, financial management, and by employing people with intellect, integrity and energy.

Wallet-size laminated cards containing the vision statement were given to every employee and are also part of the new hire packet.

Newmont's Value Statement

We, as Newmont, will:

- Demonstrate integrity, trust, and loyalty.
- Have a keen business sense, a determination to excel, and a commitment to action.
- Promote teamwork and honest communications.
- Demand positive change by continually seeking and applying best practices.
- Recognize our responsibility to our communities, safety, and the environment.

THE CHANGE ACCELERATION PROCESS

Chief Executive Officer (CEO) and Chairman of the Board Ronald C. Cambre understands and has emphasized the need to embrace change to achieve success. To help manage and encourage change, Newmont adopted a method that incorporates many of the principles developed by GE. One of the tools GE developed has been used successfully by companies around the world in a wide variety of industries. This tool, known as the Change Acceleration Process (CAP), is designed to help improve the acceptance of change. Acceptance is the method by which the creative energies of the workforce can be fully tapped. Over time, every Newmont employee will receive CAP training at least once, in many instances three or four times, and will work on one or more teams. This is the method by which GMP will achieve its goal. The CAP tool focuses on:

- Leading change
- Creating a shared need
- Shaping a vision
- Mobilizing commitment
- Making change last
- Monitoring progress
- Changing systems and structures

TEAMS

Teams are key to unleashing the creative energy of the workforce. Teams that do real work and hold themselves accountable are effective. A structured approach has been taken to ensure that each team member is clear on the team goals and roles, along with the processes for getting work done.

There are global teams with representation from each site. Among other things, these teams focus on sharing best practices and establishing standard operating procedures. There are site teams, by discipline, that focus on systematically driving out inefficiencies at the site.

("Sites" in this context can also be technical facilities, the corporate office, or even a special discipline.) In addition, "Tiger Teams" are formed to focus on a specific problem for a short period of time.

Teams are composed of staff members from different departments and disciplines to facilitate communications and make sure that any implementation does not adversely affect another department or site. For example, the site mine operations team leader is also the site representative on the mine operations global team. This breaks down the barriers between disciplines and is the first step in getting rid of the "silo" or "my site is different/special" mentality. A more "global" perspective is required to achieve the highest corporate value possible.

COMMUNICATIONS

Lack of effective communications is the root of much inefficiency. Disciplined communications sessions are one of the reasons GMP initiatives are successful. Much of GMP is about discipline. Scheduling, written agendas, starting on time, short sessions that are limited to no more than 1 hour in length, 1-page memos, and making sure everyone has the opportunity to speak are all elements of effective communication sessions.

An agenda is distributed well in advance of the meeting. A facilitator ensures that the ground rules are followed and that the meeting stays on track. A timekeeper makes sure that the meeting stays on time. A scribe documents the discussion and conclusions. The scribe completes a standard form that establishes the 3Ws (who is responsible for doing what by when) in writing. The meeting minutes and the 3W form are distributed to all participants at the end of the meeting. Again, much of this came from the principles developed by GE.

PROCESS MAPPING

A tool that consistently identifies inefficiencies is important. Process mapping visually lays out the steps in a process along with the performance measures. This helps the team to fully understand the process and to identify where improvements can be made.

Detailed process mapping is critical to the success of GMP. In many instances, process mapping is detailed down to the second. For example, in a truck/shovel operation or in a grinding mill process, action and movements would be detailed to the second.

This tool was adapted from a military process mapping method and also uses some ideas from the Lean Thinking Institute. An emphasis has been placed on how to measure the effectiveness of the process. To date thousands of process maps have been developed, covering all aspects of Newmont's business processes.

DOLLARIZATION

Dollarization is a standard methodology developed by the financial management team that is used at all Newmont sites. It is a series of techniques that link technical performance improvements to financial results. As we seek to increase production, availability, or metallurgical recovery, this helps to ensure that the potential financial impact of the results are clearly understood and widely communicated. This helps us prioritize our efforts in the areas that have the greatest potential financial value.

CRITICAL PERFORMANCE INDICATORS

A critical performance indicator (CPI) is a calculated value that supports critical business objectives. Each CPI has a baseline, a target, and an optimal value that can be compared against the actual performance to monitor progress as changes are applied.

What gets measured gets managed. As applied to GMP initiatives, this measurement component allows teams to monitor their progress in driving out inefficiencies. Based on standard definitions, these measurements allow "apples to apples" comparisons across the corporation. This establishes global standard operating procedures by identifying the best practices.

RISK MANAGEMENT

It is vital that risks be properly managed in a structured manner as teams seek ways to increase productivity and reduces costs. Risk is defined as the inherent probability for loss associated with any activity. This tool provides a structured approach to identifying, analyzing, and controlling risk.

REWARDS AND RECOGNITION

A bonus plan reinforces business objectives. When things within the control of the workers can measure the basis for the bonus, it provides an incentive to excel. Global and site rewards and recognition (R&R) teams are focused on continuous improvement of the bonus plan. With a "line of site" emphasis, each site has some latitude to select the CPIs that will form the basis from which the local component of the bonus will be paid. Again, a global R&R team is continually enhancing a pay-for-performance compensation concept.

CONCLUSION

Commitment, a sustained visible drive, and a connection between management and the employees are critical to the success of any change initiative. The communication process for these issues must be efficient with respect to every employee. This is a primary and foundational principle of GMP. GMP will engage every employee in working to drive all inefficiencies out of the company.

A standard measuring system must also be employed at every level of the organization. The measurement of CPI trends permits the financial impact of the specific efficiency issues to be identified and analyzed. Detailed analysis and evaluation, as is used in Mikel Harry and Richard Schroeder's Six Sigma initiative, is required to ensure additional process change.

Executive management, the "Officer's Group," and all department heads must remain committed, provide leadership, and constantly encourage change. The corporate level must then react to the change initiatives. At Newmont, they are accomplishing this by being connected to the teams in sponsorship and core executive roles, and by continually pursuing Newmont's vision in accord with the company's values.

Employees at every level must use the values developed by their company. Discipline must be exercised when the values are ignored. Values can be like motherhood and apple pie ... everybody has them. Any company can "talk the talk" about values, but successful companies are the ones who "walk the walk," and Newmont is "walking the walk."

SECTION 2

Company Politics

. . . .

Corporate Culture: What Is It, Why Should I Care?

Edward C. Dowling, Jr.*

ABSTRACT

Dilbert, management mumbo-jumbo, and human resource buzzwords aside, this paper defines corporate culture and why it is important to understand it. The sum of company experience, both good and bad, defines the way people work together. This integration of experience outlines the expected behaviors, or culture, in an organization. Given the competitive pressures inherent to our industry, the dizzying ongoing industry restructuring, and the recognition that behaviors drive effectiveness, the appreciation of an organization's culture is a prerequisite to meaningful improvement, particularly for those organizations undergoing growth, acquisitions, or mergers.

CORPORATE CULTURE

This paper is all about influencing behavior in the workplace and why it is important to the success of any business. It is about making positive change. Consider the controllable aspects of your job in the minerals industry. Can you quickly change the resources delivered (minerals or raw materials) to your job site? Can you quickly alter the capital equipment and infrastructure in your work area? The answer to these rhetorical questions is obviously no. In the minerals industry, the ores we extract and the processes we use to extract them are in place and we can influence only the sequence in which they are mined and delivered. Although the capital equipment and infrastructure can be defined during the construction or expansion of a mine, they remain largely intact throughout the course of the mine life. Moreover, any changes in these areas are typically very expensive, and returns on these investments can be uncertain.

Given that the mineral resource and capital infrastructure aspects of our jobs are relatively fixed, how can an individual or company make real business improvements? Experience indicates that real improvement is best made with the employees through the use of their experience and inherent knowledge capital. This situation creates a fundamental management paradox; those individuals required to make change successful must change themselves. In short, any attempt at change immediately collides with the company's culture. Therefore, a company or individual must first be able to change the environment or culture within a business in order to make real and lasting improvement.

* Cleveland-Cliffs, Inc.

WHAT IS CORPORATE CULTURE?

Corporate culture is easy to feel, but difficult to define. For example, do you feel that you have the ability to call anyone (including the chief executive officer) in your organization to obtain information to do your job better? Are you encouraged to take risks or is there a fear of failure and consequences? Are you delegated the responsibility to immediately change your organizational structure or head count, without approvals, and obtain immediate acceptance of those affected by the change? Do you think your corporation promotes a safe work environment? How do you know? At the end of the day, does everyone leaving the workplace know whether the company had a good day or not? How do they know? Can you immediately change your budget to better align with changing business circumstances? How does the company reward success? Do these rewards create more successes? Do you know your customers? How are you helping your customers meet their goals? Obviously, we can feel these circumstances, yet the definition of corporate culture remains more elusive.

With these considerations in mind, the author attempts to provide a non-Dilbert-like, common sense definition of corporate culture. Here, corporate culture is simply defined as the acceptable way people within an organization work together to get things done. The keyword within this definition is "acceptable." This implies that an inherent understanding exists within any organization of what is acceptable, or perhaps more important, what is unacceptable, when it comes to ways of working together. Generally, this institutional understanding is created over a long period of time through good and bad experiences–generally a very expensive trial and error experience. This integration of historical experience outlines the expected behaviors or culture within an organization. In short, the understood values of a corporation define its culture.

In all cases, the leadership of a corporation establishes and reinforces the work environment, which, over a period of time, defines its accepted behaviors or culture. These accepted ways to work together are reinforced by the systems and structures of a company. A few examples of systems and structures include the company policies for:

- Resource allocation connecting strategy to budgeting
- Organizational design and structure
- Hiring, placing, and retaining talent
- Building competence and capability through training and development
- Effective communication of the right information to the right employees
- Use of technology and sharing of best practices
- Measuring business and individual performance
- Compensation and rewards

There is an old saying that demonstrates the importance of systems and structures: "*What is important, gets measured–what is measured, gets rewarded–what is rewarded,* ***gets done!***" Of course, people and companies that get things done get ahead. Unfortunately, changing the policies about a company's systems and structures can be very difficult. Changing systems and structures is all about obtaining alignment in the organization. In some cases, the corporation will deliberately establish its systems and structures to reinforce its company values and strategy. All too often, however, a company's systems and structures do not support the company's direction. This makes it very difficult for a company to achieve its desired outcomes, simply because the organization is not aligned for success.

WHY IS CORPORATE CULTURE IMPORTANT—WHY SHOULD I CARE?

We are living in a world of dizzying and accelerating change. Many forces are causing this change: globalization, the technology revolution, social expectations, and regulation, to name a few. This affects us in ways we can see around us everyday. Think about the equity value decline of the minerals industry over the last 3 years. Simply stated, the investor community does not value the minerals industry for an expectation of future earnings. Within a business, this is why we would have to use scientific notation to count the number of times we have heard the concerns about costs. On the strategic level, this is why we have seen the acceleration in the number of mergers and acquisitions in the last years. Think about the changes within your immediate business in the last year. When was the last time your boss discussed head count with you? The conclusion here is that the need for fundamental change and meaningful improvement is required for the minerals industry to prosper and that the forces of change will continue to accelerate.

The role of management is to lead this change, yet the existing culture within an organization may preclude effective change in support of sustaining the status quo. Other corporate cultures promote achievement. What are the differences between "sustainers" and "achievers"? A simple description of the sustainer culture is one in which the environment is such that the workers believe that all decisions are treated and scored by the company in a pass-or-fail manner and that there are severe negative consequences for a poor decision or a failure. As such, the organization avoids making new decisions because of fear of the consequences of failure. The values of trust and confidence are very low in sustaining-type organizations. On the other hand, achieving organizations are described as those where important decisions are not treated as pass-or-fail exercises; instead, these change decisions are treated more as experiments. As such, the company is willing to tolerate well-meaning mistakes in pursuit of business improvement. Individuals within these companies are encouraged to think, share ideas, work together, take risks, and learn. Over a period of time, these achiever-type companies learn important lessons from both failures and successes and become very adept at making change and improvement. This is why having the "right environment" is always cited as a prerequisite for innovation in any organization.

A fundamental truth in business is that people or companies that get the **right things done** get ahead. This implies that to identify the "right things," there exists a good base for connecting business strategy with technical opportunities. Generally, most engineers and companies are good with this sort of planning. However, the ability to effectively implement and deliver the business results is where most organizations fail. As previously discussed, these technical initiatives (or changes) collide head-on with the corporate culture, and, in most cases, the culture wins. Hence, it takes too long to implement the change or it is implemented inconsistently, and business results are less than optimal. As a result, the resources dedicated to the change are wasted and value is destroyed.

Fortunately, there is a better way to implement *change*. With this methodology, a technical strategy is integrated to an organizational influence strategy to build acceptance for change. In addition to developing a technical solution for change and improvement, a corresponding strategy must also be developed and deployed to overcome the resistance of the corporate culture to change. Simply stated, this acceptance strategy must create a force for changes that can overcome the resistance to change. Successful companies apply a variety of techniques to overcome the resistance factor of a corporate culture. Cleveland-Cliffs and other minerals companies have adopted the Change Acceleration Process (CAP) to assist cultural change and improvement. This tool, developed by General Electric, provides simple, straightforward methods to organize a change project and integrate a cultural strategy to a technical strategy to accelerate meaningful business results. The basic tenet is that broad organizational involvement is required to overcome a technically and culturally complex problem. Experience indicates that these difficult problems hold the greatest potential for improvement.

CAP employs a simple formula:

$$Q \times A = E \qquad \text{(EQ 1)}$$

Where:

Q = the quality of the technical solution
A = the acceptance of the technical solution
E = the effectiveness of the solution, or the business result.

The power of this concept can be demonstrated using a scale of 1 to 10. An organization can have an excellent technical solution (say 8), but a poor acceptance (say 2) and arrive at a sub-optimal business result (16). As engineers, we generally retreat to the drawing board at this point to improve the business result and come back after spending more resources to increase the technical solution (say 9). In this case, the overall business result increases marginally (18). However, if the organization acceptance is improved (say 8), a much more meaningful business result is created (64). Please recognize that although CAP has proven successful, many other models exist to overcome cultural resistance. The real value comes from successful application of these techniques to solve problems and realize business benefits.

Although these concepts and tools are extraordinarily useful within any business, they take on particular value for those companies dealing with the cultural issues arising from mergers and acquisitions. In these situations, two or more companies, having their own independent cultures and different ways of working, are thrust together and expected by their management and investors to extract the value anticipated by the business transaction. More often than not, these mergers fail to extract the anticipated values within the desired time frames simply because effective company integration is precluded by cultural barriers.

SUMMARY

Corporate culture can work for you or against you. A major role of management within any company is to lead change within the organization. Given the current business environment and economic forces, the requirement for this capability is likely to increase. It is those individuals and companies who embrace change and can lead change by molding their corporate culture to the changing business circumstance that will survive and prosper. By coupling a cultural influence strategy to a cultural strategy, change can be a very positive and rewarding experience. With experience and positive results, these organizations obtain a special capability to react quickly to their customers and create extraordinary value. On the other hand, those organizations or individuals that cannot adapt or lead change will, over a period of time, fail, and others will capture their resources and inherent value.

. . . .

Will It Be Business As Usual?

Comments and Observations on Mergers and Acquisitions in the Mining Industry

Fred H. Lightner*

ABSTRACT

Mergers and acquisitions (M&A) within the mining industry are becoming frequent events. Large companies and small companies alike are jumping on the bandwagon of merger mania. The results of these mergers have significant impacts on the employees of the involved companies. Comments, observations, and specific examples of the "people" aspect of mergers are presented in an attempt to answer the question "will it be business as usual?"

My answer is a simple ***NO****. This paper attempts to provide support for my conclusion.*

The reader should recognize that the comments, observations, and specific examples mentioned in this paper are subjective and reflect the experience and prejudice of the author.

INTRODUCTION

Mergers and acquisitions (M&A) within the mining industry are becoming frequent events. In 1999, the number of mining-related mergers in excess of U.S. $10 million increased to 100—the highest number since Raw Materials Group started its annual M&A review in 1995. Roskill Information Services of London publishes "Who Owns Who in Mining" (*The Northern Miner* 2000). Large companies and small companies alike are jumping on the bandwagon of merger mania. These mergers have significant impacts, both positive and negative, on the employees of the involved companies. From the individual employee perspective when a merger occurs, "will it be business as usual?"

My answer is a simple **NO.**

COMMON CORPORATE MERGER AND ACQUISITION TRANSACTIONS

For general reference, a brief summary of the more common types of corporate M&A transactions is presented. The transactions can be described as:

* Metallica Resources, Inc.

Merger of Equals

A merger of equals combines Company A and Company B of similar size into one "NEWCO." Ideally, the ownership of NEWCO after the merger would be equally distributed with the owners of Company A and Company B each owning 50%.

Acquisition or Takeover

An acquisition or takeover occurs when Company A purchases Company B. Company A continues to exist and operate; Company B no longer exists.

Reverse Takeover

A reverse takeover occurs when Company A purchases Company B, but the owners of Company B wind up with a large enough ownership position in Company A to actually control it.

Asset Purchase

In an asset purchase, Company A buys a specific asset, such as an operating mine or exploration property. The fate of any employees who may be operating or managing the asset is usually negotiated as part of the purchase.

Any of the above transactions will have an impact on the employees of the involved companies. For the remainder of this paper, the term "merger" will be used as a general term to refer to any of the above transactions. The term "NEWCO" refers to a hypothetical recently merged entity.

AFTER THE MERGER—INITIAL SURVIVAL

One of the most important and most evident potential impacts on employees involved in a merger is the first question: "Do I still have a job?" Mergers often tout the potential savings in corporate overheads by combining two companies. Although there are nonpeople costs included in the savings, one should realize that most of the time some employees will stay and some will go. It is natural for a recently merged company to want to show the benefits of the transaction, and work force reductions are sometimes made to demonstrate the efficiencies and cost savings of the merger.

In an industry that is consolidating and with employee contracts becoming more commonplace, at times there are strong incentives for people to leave. Who gets to stay and who has to go can become reversed to who has to stay and who gets to go.

ADJUSTING TO THE MERGER—BUSINESS AS USUAL FACTORS

If they survive the merger, employees must then adjust to NEWCO. Below, in outline form, is a checklist of business as usual factors. These items affect the individual employee, the everyday work environment, and how the employee functions on the job.

Outline of Business As Usual Factors

- Company profile
  - Size and scope of activities
    - Market capitalization or ownership
    - Amount of debt
    - Capital spending and budget
    - Production levels

        - Number of employees
        - Earnings and cash flows
        - Commodities produced
        - Number of mines, operations, offices
        - Geographical focus
        - Exploration or acquisition oriented
    - Operational style
        - Conservative or aggressive
        - Proactive or reactive
        - Promotional or low profile
        - In-house or outsourced functions
- Company policies and politics
    - Organization
        - Is there an organization chart?
        - Who is my boss?
        - What is my job?
        - What is my authority and responsibility?
        - Who are my peers?
    - Employment policy
        - Employment contract
        - Relocation
        - Cash compensation
        - Bonus
        - Stock options or profit sharing
        - Benefits
        - Vacation
        - Raises
        - Performance reviews
        - Opportunity for advancement
        - Training and professional advancement
    - Company policy
        - Travel
        - Expenses
        - Dress code
        - Budgeting
        - Purchasing
- Management style
    - Micromanagement or delegation
    - Head office or local control
    - Communications

- Required reports
- Distribution of reports
- Frequency and purpose of meetings
- Formal or informal
- Inside or outside the chain of command

– Sacred cows

- Technology, such as computers and software
- Suppliers
- Certain employees within NEWCO
- Consultants or contractors

The most fundamental point to realize about M&A is the golden rule–he who has the gold makes the rules. Simply stated, the majority owner or the control block will directly or indirectly manage and operate the company. Regardless of whether a company is publicly or private owned, someone or some group is in control. Even senior management and boards of directors are not immune from majority owners. From the individual perspective, if you are an employee of NEWCO and the majority ownership changes, you should expect change. Decision makers are never identical and they will tend to manage from their individual bias and experiences. They will tend to use the people, methods, and policies that they have determined to be most effective for them.

Even if majority ownership does not change in NEWCO, the individual employee should expect some change. To assimilate the newly acquired company or asset of NEWCO, changes in management styles, company policies, and company politics may be required. In either case, do not expect business as usual.

SEE WHAT I MEAN?

Here are four examples that are based on actual events associated with mergers and company politics. Individuals and companies will remain anonymous.

Example 1—Company Policy

Company A, a large oil conglomerate, acquires Company B, a smaller oil company with a minerals group. John Gungho is the project manager for Company B. John is in the midst of developing a small precious metal mine that is being built in the western United States. Several of the employees of both Company A and B are relaxing with a cold one after one of the get-acquainted "dog and pony shows." An employee of Company A asks John, "What is the largest invoice that you have approved for payment?" John casually replies, "$2 million." The employee of Company A states, "Gosh, even Tom Bigwig, the executive vice president of Company A (the boss of John Gungho's boss) can't do that!"

Example 2—Company Politics

Two young engineers, George Miner and Herman Metallurgist, are attending their first national meeting of the American Institute of Mining, Metallurgical, and Petroleum Engineers (AIME). Both George and Herman work for the same company but one is located in the Eastern Mining Department while the other is in the Western Mining Department. The company is hosting an employee cocktail party before the banquet. George is anxious to inform Herman that according to his boss the official dress for the company party is a white shirt and a black skinny tie. Herman checks with his boss who says nonsense. Being somewhat of a rebel, Herman attends the party in a black shirt and a wide white tie.

Example 3—Sacred Cows

Company B has a team of employees and consultants reviewing bid proposals for the feasibility study of a large Latin American mining project. After exhaustive analyses the team submits a detailed evaluation of the proposals, along with its recommendation to select Engineering Company X. Notifications are made and a kickoff planning meeting is held. Within a few days, Company A takes over Company B. Company A terminates the agreement with Engineering Company X and engages Engineering Company Y. The rationale for the change is simple; Company A always uses Engineering Company Y.

Example 4—Always the Survivor

Company A, a small precious metal company, has one operating mine and a very small corporate office in the western United States. Through a series of acquisitions and financings, Company A grows to a multimine, multinational producer. Several management changes, precipitated by the company growth, are made along the way. Although Company A was always the survivor, the company profile, operating and management style, and politics evolved into something completely different than its beginnings.

SUMMARY

Almost everyone associated with the mining industry has either had some personal exposure to M&A or knows someone who has. The purpose of this paper is not to throw stones at the merger mania taking place; in many instances, the consolidation makes sound business sense. However, employees who may be affected by M&A should realize that changes will happen. It will not be business as usual. Hopefully this paper will provide some insight to assist with the personal adjustments required when there is a change in corporate ownership or control.

REFERENCE

Roskill Information Services of London. 2000. Commentary, charting merger trends. *The Northern Miner* 86(22):4.

Project Handoff from Exploration to Development: How to Avoid Fumbling the Ball

Gregory A. Hahn*

ABSTRACT

Explorationists and engineers rarely speak the same language. Explorationists are not cognizant of the information that engineers require to make the appropriate choices and decisions. Engineers cannot "translate" geologic parlance and data into information relevant to proper decision making. This often creates a physical and intellectual "rift" between engineers and geologists that inhibits a smooth transition to development. A transition team leader who is grounded in both disciplines can best facilitate integration and communications between the explorationists, who understand the intrinsic attributes of a given deposit, and the engineers, who have to design for extraction and beneficiation of an orebody.

INTRODUCTION

My personal experience in predevelopment projects began in 1978 when I was assigned as project geologist to the development team assembled by Noranda, Inc., for the development of the Blackbird cobalt-copper deposit in central Idaho. This project failed to mature into an operating mine for various reasons, as did many of Noranda's predevelopment and development projects during the 1970s and 1980s. Certainly one contributing factor was the awkward transition from exploration project management to development (and the conflicting objectives and management styles of the two efforts). After the frustrations of our predevelopment failures at Noranda, I joined CoCa Mines in 1986 as chief geological engineer. My task was to take control of predevelopment for the small precious metal producer, having developed a strong conviction for how and how not to bring about this critical stage in a project's evolution. The title of that position reflects in part the philosophy I developed on the subject. In the 5 years I was with CoCa Mines, we successfully brought four projects through predevelopment and two into production.

CoCa Mines was sold to Hecla Mining Company in 1991, and I joined a small base-metal development company with the mission of bringing late-stage exploration projects through

* President and CEO, Summo Minerals Corporation

predevelopment and into production to build an operating company around the property assets. We are currently engaged in the latter as we build Summo Minerals Corporation, a Toronto Stock Exchange-listed company, into a mid-tier copper producer. In 1995, we hired a general manager/vice president of operations–a metallurgist/process engineer–to assume management of the Summo's primary asset for permitting and construction, and ultimately for operations. The challenges we faced included effecting a smooth handoff from advanced exploration to development, as well as transferring the geologic information critical to mine development to the engineering team for permitting, ore reserve calculations, and mine and process design, while avoiding fumbles that delay or cripple a project's development and could possibly jeopardize the project's survival.

WHAT MAKES SMOOTH TRANSITIONS DIFFICULT

The transition of a project from exploration and research to development entails a major shift in project goals, interim objectives, project personnel, and management styles.

The Birth of a Project

Colleagues of mine have drawn analogies between project transitions and the birthing process. The conception of an idea and the nurturing of that idea involve a certain degree of privacy and intimacy in an atmosphere sheltered from outside influences and forces. This environment is critical for protecting the fledgling idea while it is still underdeveloped and vulnerable. The idea is generally mothered by a small close cadre of individuals or one individual, called the project champion, who truly treats the idea (project) as their own offspring. Without this level of dedication and devotion, the idea has little chance of surviving. Once the idea demonstrates merit, or is born, it is by necessity exposed to many more influences than while it is in the protective womb of its progenitor. Just as a newborn is exposed to light, air, cold, viruses, bacteria, clothing, and new foods, among others, a new project needs to undergo a plethora of new experiences to develop the mettle it needs to mature and survive.

The manner in which we expose the project to these new experiences determines its future viability. Just as a newborn needs to be progressively introduced to new experiences, a project must undergo a smooth transition into its brave new world. We rarely consider separating a parent from a newborn solely for the benefit of the newborn. Yet we often–in fact regularly–sever the progenitor from the new project under the guise of the project's best interest. When we commit this error, we sever the project from the person who knows the most about the project's background and history, its subtle strengths and weaknesses, and its likely response to various circumstances.

Other Influences

In addition to the birthing-parenting analogy, several other aspects of the standard project transition can result in an awkward or erratic development of a project. I consider the principal among these to be:

- Language and communication barriers
- Pride of authorship
- Ignorance of the project's past and future
- Poor comprehension of the entire project scope

I'll offer some comments and observations on each of these issues based on my own experiences in attempting to bring mining exploration projects through predevelopment and construction.

Language and Communication Barriers

In mining, the technical knowledge and the language used to communicate that knowledge are not necessarily compatible between the exploration–research group and the development group. This is probably true of most technical fields. In mining, the exploration group is composed primarily of geologists, geochemists, and geophysicists, who even among themselves often have difficulty communicating and understanding each other's disciplines. The topics and issues that are important to them in understanding and nurturing their concept or project may not be important to the development team, which is composed of mining engineers, metallurgists, environmental specialists, and financial experts. This is not to imply that they are unimportant, but because of language barriers and a lack of understanding between disciplines, it is often difficult or impossible to identify and communicate the important elements.

Communication of critical project parameters becomes even more difficult if the original project team members (the nurturers) are excluded as the project advances through development. It is presumptuous to believe that all critical elements of a project will be documented in writing, or that if documented, they will be referenced and understood. I've witnessed a geologist who retrieved a dust-covered report from a file and pointed to page 100 and something, exclaiming "I said right here in this report that this portion of the deposit contained elevated mercury...." That is not an effective means of communication, and does not lead to successful development of a project.

Pride of Authorship

Despite the fact that all individuals involved in work on a project, both preceding and during the transition, are professionals, we must recognize that personalities, as well as professional and personal pride, do play a role in the ease or difficulty of transitioning a project out of exploration and research. It is easy to understand the pride the nurturers have in the evolution of their idea or project, and the injury their pride may suffer as they watch their project taken over by a team of strangers who have entirely different values, if not even disdain for the nurturers' own values. It is also easy to appreciate the development team's frustration when they are unable to identify information they deem critical to advancement of the project among the voluminous data the nurturers have presented, and their need to develop their own sense of ownership, which is required before they are fully committed to the project's future.

This lack of understanding and respect between the two groups (and the different, yet essential, roles both play in the successful evolution of a project), is perhaps the most significant contribution to friction between the two. This deficit in understanding is one reason that vital information is not adequately transferred in the transition process.

Ignorance of the Project's Past and Future

One of the fates of a new project within the business world is the need to fast track it once its merit has been demonstrated, in order to maximize its impact on the market and its profitability (or return on investment). This often entails a change in pace for the project, which is not necessarily in accordance with the project's historical pace. This is especially common in project development outside the United States and Canada.

The need arises to proceed on to tasks D, E, and F, which logically build on tasks A, B, and C. However, often the initial tasks have not been completed (or worse, have not yet begun). In this case, the downstream work on the project may be built on a weak foundation. This often happens when nurturers have an inadequate understanding of where their project may ultimately be heading. Conversely, a project may proceed to downstream tasks without full knowledge of the history of the initial tasks and the starts and stops the project may have experienced or suffered along the way. This may happen during property purchases, mergers, takeovers, or when there is a general change of project personnel. Any of these can damage a project's potential and future if unattended.

Poor Comprehension of the Entire Project Scope

An expansion on the lack of continuity mentioned above is the lack of overall understanding of the magnitude of a project, the components required to make a successful project, and the steps or paths that need to be covered to bring a new project to construction. It is often assumed that all the geological, geochemical, and geophysical data–and only these types of data–are generated in the exploration-research conception of a project. The engineering designs and subsequent data are then built and generated in the development stage based on the existing geological knowledge. A grievous error is made when the two sets of data are not allowed (or encouraged) to be developed simultaneously, so that they can build on and support each other.

Most often the exploration-research group is unfamiliar with the geological needs of the downstream users. Similarly, the engineers are often unaware of the assistance that the nurturers could provide in developing data specific to their needs. As a result, the two groups develop two sets of data that often are not compatible. Much vital information gets lost or falls into the gap created by this approach.

RECOMMENDATIONS

I have found several approaches to be useful in facilitating a smooth and effective transition for a project from exploration-research to development. They stem from acknowledging the problems outlined above and represent an attempt to cover or accommodate some of the shortcomings experienced in project transitions in the past. As soon as a project is recognized as having potential for development, an engineer needs to be included in continued work on the project. This should occur early enough in a project to allow the engineer an opportunity to express the likely downstream needs of the future development team, and to help develop the requisite data.

Toward this goal, it is important for a company to develop an environment where explorationists and engineers rub shoulders and are encouraged to communicate, work, and socialize with each other. If the company cannot afford to have an engineer on staff to work with exploration, the company should draw on resources from the consulting community as the need arises. The goal is to develop a relationship with an engineer or engineers that promotes familiarity and comfort among the respective players. The engineer who participates in the nurturing process can be a significant advocate for the project (pride of authorship) with the future development team, and an effective translator for the two teams in transferring information from one to the other (language and communication). The engineer, in becoming the bridge between the gap that exists naturally between the geologic and engineering disciplines, must feel a part of both entities in order to be an effective representative for the project.

Clearly the engineer's participation must continue from advanced exploration through the transition period. Similarly, at least one representative of the exploration-research group that nurtured the project should be invited or required to participate on the development team. Management of the development team clearly needs to evolve to the engineers, but the engineers need to be aware (or made aware!) that the success or failure of the project rides on a proper understanding of the geology. Mining, processing, and environmental considerations all are rooted in the fundamental geology of the project.

As engineering and environmental issues arise, a geologist familiar with the details and history of the project must be available to help address these issues. Both geological and engineering professionals need to accept and appreciate that both groups play a critical role in the successful development of a mine or company. It is management's responsibility to create an appropriate atmosphere where these two groups come to know and respect each other, and are thus able to work together to achieve the mutual goal of bringing a new mine, well, plant, or product on the market.

My experience and conversations with other professionals indicates that these objectives are more easily accomplished in smaller organizations than in larger ones. I suspect that this derives from the much simpler and more streamlined organization inherent in smaller companies and the tighter sense of community and team spirit present in the smaller organization. It is easier for all personnel to feel that they are contributing and that their contribution can make a difference when they don't appear to get lost in the organization chart or the employee directory. Teamwork is the key to success, and the team should be built on the strengths of both the geologists who understand the intrinsic attributes of a given deposit, and the engineers who must design for extraction and beneficiation of an orebody.

John Doe Bloopers, Blunders, and Scams

William F. Riggs*

ABSTRACT

The name of this symposium speaks for itself in the connotation normally assigned to the term "political." In this paper, various donors have contributed their stories of metallurgical bloopers, blunders, and scams. The paper covers areas from human error stories, through technical blunders, to outright scams. Included are some stories that involve geology as it relates to metallurgy, warehouses of precious metal concentrates, reagent selections, solutions that treat the symptoms instead of the problems, improper technological transfers, humorous design mistakes in the interest of being exacting, and many more. The stories are all true with the names and dates withheld and the cases presented in the true sense of the "John Doe Stories."

INTRODUCTION

One of our fine members of the mining profession once had the honor of becoming the President of the United States of America. His name was Herbert Hoover. He was given credit for making a very profound statement as the keynote speaker at one of the American Institute of Mining Engineers' annual meetings during the 1950s. Allegedly, he said "our mining industry manages to survive in spite of ourselves!" This paper is presented to describe some of the events and happenings that show that this has some basis in truth. The examples presented here were supplied by a host of members of the mineral processing industry. Not all the examples could be included in this paper because of restrictions of time and space. Some of the examples considered for inclusion were a sad demonstration of the state of our industry's performance; others were so hilarious as to result in comical disbelief and laughter for hours afterward. The decision was made to present those examples that basically maintained a middle road and would provide some guidance and warnings for future operators as opposed to excessive humor or negativity.

* CYTEC

Case Number 1

Employment practices have often been influenced by events that have lasting effects to the point that the innocent can suffer the sins of those who follow after them. For example, a certain mining company would not hire mining graduates from a particular university for a period of 6 to 8 years. The company had had an option on a mineral prospect, and they had sent a young geologist to supervise the drilling program and evaluate the core drilling for values. After the drilling program was complete, the young geologist sent back word that the program had encountered only a small seam of massive pyrite and that no minerals of value were found. The company dropped the option on the prospect. At a later date, budgetary members from the head office made a cursory visit to evaluate how the money was spent. The resource turned out to be a 4-foot seam of massive chalcopyrite mineralization! The company then had to renew the option on the property at 4 times the value of the original option because the landowner knew that the company would not be returning after leaving without good reasons. After that experience, the young geologist was requested to seek employment elsewhere and no one in the company would risk hiring anyone from that school for about 8 years.

Case Number 2

A group of foreign investors wanted to build a hotel in a South Pacific island paradise. They could not obtain funding from the banking community unless they came up with 40% of the capitalization. They proceeded to find private investors by claiming that they had 100 barrels of precious metal concentrates in a warehouse in the western United States. They had a certified analysis from a U.S. laboratory, as well as pictures supporting their claim of more than 4,000 ounces of platinum group metals per barrel. Several investors were interested, especially when one does the math (4,000 ounces multiplied by $600 per ounce times 100 barrels).

When one of the investors had a metallurgical consulting firm investigate, however, the barrels were filled with concentrated alluvial fill of no value. They did have some good pictures to show the investors in the beginning to stimulate their interest. It never hurts to read the fine print!

Case Number 3

A group of very successful real estate investors from one of the tourist islands in the Atlantic Ocean had been approached by an inventor who had developed a method that used flotation to recover platinum group minerals in the United States. The inventor successfully demonstrated the process in his laboratory. With the results of the laboratory tests, repeated many times for those real estate investors who had doubts, he convinced them to build a concentrator next to a deposit he had discovered near a metropolitan city in North America. The investors, greedy to increase their recent real estate windfall, could not furnish him enough money fast enough in their desire to get into production. When this project was 80% complete, one investor decided to bring in a consulting geologist and metallurgist to confirm the results (a little late, maybe?). The geologist looked at the deposit, a major limestone reef, and declared that no mineral values had ever been found in this formation. When the metallurgist with the secret process became incensed, the consulting metallurgist took some ore and ran the process exactly as the inventor had described. After sending it out for analysis, the project was dropped. However, the investors had already spent more than $5,000,000 in capitalization, on which the inventor had taken 10% off the top.

Case Number 4

There always seem to be tales of how we in the mining industry continually treat the symptoms of a problem instead of solving the problem. This case illustrates how we continue to do this and also lends veracity to President Hoover's alleged statement.

A fluorite/lead/zinc operation was not making grade. Operators continued to add cleaner cells to the flotation circuit to improve the grade of the concentrates. The cleaner flotation circuit circulating load thus increased to the point that they had to dump the cleaner tails from time to time, resulting in reduced recovery. Their solution was to add even more cleaner flotation cells to add flotation time to the cleaners for the purpose of making grade while not having to dump the cleaner tails as the circulating load increased. That failed, and the problem persisted. Next, someone brought in a mineralogist and he determined that they could never make grade because there was a high percentage of locked particles that made up the circulating load.

Why had they not determined this before? Because over time, they had made small incremental changes to the grinding circuit. Each one did not appear to be important in affecting the final performance of both grade and recovery. However, the cumulative effect was to cause a significant reduction in the grinding efficiency to the point that it affected the final product. In other words, the initial design was working well, and they did not think that each incremental change would have an effect. They failed to take into consideration the cumulative effect as it slowly developed.

Case Number 5

This is one of my favorite stories, which illustrates the need to cross train mineral processing personnel. A fluorite mine and mill was operating at fairly large tonnage for a nonmetallic operation of this nature. The mine and mill had its headquarters on the other side of the continent and had sent its experienced engineers out to start up this state-of-the-art operation. It was potentially very profitable because it was during the time that acid-grade fluorite concentrates from flotation were bringing a premium price. After starting up the mine and float plant, the operation began to function as they had predicted, and was very profitable. The company then gave permission to employ mine and mineral processing engineers in that region so that the corporate technical team could return home. Meanwhile, copper prices had dropped significantly (which they do every 8 to 10 years whether they need to or not), and the result was the availability of many experienced copper flotation engineers. The engineers were given a brief orientation period and allowed to take charge of the operations; the corporate experts returned home.

Approximately 6 months later the grade went bad. They could no longer make the premium acid-grade fluorite on which the economics of the plant was based. Telephone conferences went back and forth. The reagent dosages had not changed, the flowsheet was the same, a change in the mineralogy could not be determined, and the pH was the same. The experts in the home office could not understand what was happening, yet *something* had gone wrong! Finally the company sent out two experienced fluorite flotation engineers. They started going through the procedures one step at a time. By the time they reached the pH, they were about to pass over it because nothing was out of line. Then one of the copper engineers pointed out that not only had they maintained a good pH control, it was even now at a lower cost after they had switched from sodium carbonate to lime. Being copper flotation engineers, they had not been trained in the surface modification of minerals by the interference or adsorption of soluble cationic metallic ions. The nonmetallic chemistry of flotation did not simulate metallic flotation at all, but they had no way of knowing that the resultant surface chemistry of nonmetallic flotation was much more sensitive.

Case Number 6

A coal tailing reprocessing plant was designed and built by an engineering company that did not have many experienced operations personnel. The plant would make a good grade of coal when operating at approximately 80% of design. However, each time the plant reached design tonnage, the grade would eventually drop via an increase in ash or silica. The problem was to increase the screens and cleaning capacity by 30%. The problem continued to persist without solution. Finally, two experienced operators were called in to audit the flowsheet. They discovered that the young inexperienced engineers had installed an overflow spout on the hydrocyclone slime's overflow sump. In itself, this was not a problem; however, in order to keep the floor clean the overflow spout from the slime sump returned to the hydrocyclone feed sump! Every time the plant reached tonnage, there was no spill to indicate a volume problem because the excessive slimes being removed were returned to the feed to the plant and the slime content would build up like an excessive recirculating load. (A good analogy would be the recycling of biological waste matter from a pigpen back into the pig feed.)

Case Number 7

One case in point shows that trying to develop a harmonious plant operation can be taken too seriously. A phosphate plant was constructed in the southern United States. The process involved considerable screening and washing of the phosphate pebble. When the plant started up, it was truly harmonious. All the screens vibrated in the same direction at the same speed. However, within a few hours of startup, when all the screens were in operation, the entire plant began to sway back and forth! The plant had to be shut down immediately or it would have come tumbling down like a house of cards.

Case Number 8

At one time, a reagent called ammonium sulfide was used in the mining industry. (For those who have never smelled ammonium sulfide in concentrated form, congratulations!)

Ammonium sulfide was the reagent of choice in pretreating molybdenum–copper flotation concentrate before separating the molybdenum mineral from the copper mineral. The mill that decided to use this technology was in the far northern part of North America. As a result, the mill was closed up very tight during the winter to avoid the ravages of the cold weather. All the bulk storage tanks were inside the plant in the basement. The ammonium sulfide was placed into a new 30,000-liter storage tank in the basement of the concentrator. When the application started, the process was working very well, and it was declared a success. However, after several days, the odor became worse and worse. Investigation of the connections at the bottom of the tank found that the specified steel valves had been ignored and brass valves had been installed because safety personnel determined that brass was safer (because of its reduced tendency to spark around the chemicals). The problem? They had failed to check with the supplier or review the chemical reactions of the product they were storing. As they learned the hard way, ammonium sulfide is an excellent solvent over time for any brass components!

The result was that the ammonium sulfide had to be rapidly pumped out into the tailing to remove it from the plant. Ammonium sulfide smells like a combination of concentrated smelling salts and rotten eggs, and I do mean "concentrated."

Case Number 9

We are a congenial group. We take in all disciplines from other industries and welcome them to our fold. However, when we do not police our own practices, we can have some interesting results. For example, some petroleum engineers were assigned to design a mineral processing plant. Their inexperience resulted in designing slurry lines that feed the hydrocyclones that contained 5- by 90-degree turns, 4 flats, and 2 verticals for 125 feet.

When the plant started up, all lines sanded up as a result of a 30% drop in velocity to the hydrocyclone. As a result, they could not distinguish between slurry and liquid.

CONCLUSION

The purpose of relating these scams, bloopers, and blunders is to show that it's up to *us* to make sure that we don't just survive in spite of ourselves. We must be better than that. We must combine maturity, objectivity, and experience if we are to function as the professionals we are supposed to be. We must use our common sense and draw on our experience to avoid in the future the blunders, bloopers, and scams that have been so prevalent in our past.

. . . .

The Business Unit Model for Mining Companies

Bruce D. Hansen*

ABSTRACT

The complexity of large geographically diverse or multicommodity mining companies requires a corporate structure, which provides the maximum long-term effectiveness. Structure models exist along the spectrum of centralization to decentralization. This paper examines the trade-offs of flexibility, cost, decision making, consistent strategies, and corporate control associated with different structural models as applied to mining companies. The mining industry typically lends itself to a business unit model. However, the degree of decentralization depends on size, proximal operations, technical complexity, product mix, geographic spread, mining methods employed, and the inherent capabilities of an organization.

DIFFERENT ORGANIZATIONAL MODELS

Mining companies have organizational structures similar to those found in general industry. They range through the spectrum of highly centralized (i.e., command and control) to highly decentralized (i.e., a financial holding company). The following sections illustrate four different models that span this spectrum.

Financial Holding Companies (Model A)

Financial holding companies are the most decentralized. They are, in essence, financial investment companies that influence their strategic business units (SBUs) or held companies through board involvement or by financial directives to the SBU management team. However, examples of financial holding companies in today's mining world are rare, although early in Newmont's corporate history it could be considered a financial holding company. (More discussion on Newmont's structural evolution is provided in Section III.)

Business Unit Model (Model B)

In this model, the SBUs are relatively self-contained with core corporate functions providing strategic direction, policies and procedures, business unit goals, and general oversight. The corporate core also raises financing and manages the corporate image.

* Newmont Mining Corporation

Business Units with Active Corporate Staff Involvement (Model C)

The corporate core is typically larger in this model as compared to Model B. Corporate staff is more involved in supplying services, support, and expertise to the SBUs and, typically, the corporate staff is involved with major site decisions. Generally, the corporate core provides both strategic leadership and consulting services. This is the most dominant model employed by large mining companies.

Command and Control (Model D)

In this model, which is the most centralized, decisions are predominantly made at corporate headquarters (top-down), and the corporate core acts as the key managers of the operating sites.

Most mining companies utilize one of these models, depending on their (1) corporate history and culture, (2) product diversity, (3) geographic diversity, (4) technical complexity, (5) organizational capabilities, and (6) size.

By their very nature, mines can be considered a form of discrete business unit, just like plants or factories in other industries. Mines may not, however, be SBUs in the traditional sense, depending on the organization model applied by the corporation. Mining companies may have SBUs created by product (i.e., copper division, aluminum division, industrial minerals division, or gold division, for example) or by geographic area (i.e., North American operations or African operations) Some companies, especially those that are more centralized and often smaller, just may have mines that are not SBUs.

Corporate history and culture have an important impact on where mining companies lie along the decentralization to centralization spectrum. Small companies or a company with a few small mines tend to be more centralized, as the small mines don't have the financial mass or need more sophisticated talent sets. The corporate office in these cases tends to provide more support and oversight. On the other hand, some small companies with a single large operating site will concentrate their talent at that specific site. Stillwater Mining is a current example of this approach, as was Getchell Gold before its acquisition by Placer Dome. In each instance, these companies maintain or maintained very small corporate staffs, focusing on core corporate functions, such as the office of the chief executive officer (CEO), investor relations, accounting, and taxes. Even typical corporate functions such as human resources and information systems are handled at the single large site.

The cultural evolution of a particular company can also dictate the organizational model employed. Entrepreneurial organizations that have been directed by a strong CEO and his core team will tend to be operationally focused and will often be more centralized, even as the company grows larger. The core team that has built and operated the company from the beginning often finds it difficult to delegate responsibility and authority. Companies that have grown by acquiring mines or other companies may tend to be less centralized.

If a company is concentrated in a single geographic region but has a number of small mines, the tendency is to centralize, especially technical talent, to gain economy of scale. Companies that have grown to conduct large operations in disparate regions, even internationally, will be less centralized, and it will tend to create separated, regionally focused SBUs. Often, from a management perspective, it is very difficult to control geographically widespread operations from headquarters. Even in a world with modern telecommunications technology, different time zones and different political and cultural environments dictate that sites operate more independently. Also, companies that produce different commodities will tend to operate in a more decentralized fashion with separate SBUs for different product commodities.

From a technical management standpoint, we often see companies that maintain a strong headquarters technical group, which acts as internal consultants and mine developers. This core group gains experience by working on the development and operational optimization of many mines. This group also serves as a valuable source of internal institutionalized knowledge that solves site technical issues and directs long-term strategic research and development efforts. In the gold business, most large companies still retain a significant centralized technical staff. Examples include AngloGold, Newmont, and Placer Dome.

The models employed by mining companies vary depending on their unique circumstances, capabilities, and needs. Obviously, larger more diverse mining companies have more options, but there is no one-size-fits-all model. Other factors that can influence the management model are:

- Linkage between sites or SBUs–The less overlap of customers, products, resources, and risks, the more autonomously the units should be run.
- Market Dynamics–If a company's units are so intertwined that decisions in one unit or SBU have an impact on the performance of the other units, it is difficult to manage the units separately. Newmont's Nevada operations are cross-dependent in this regard.
- Style/Culture–A hands-off leader will tend to delegate more; a nuts and bolts, operationally focused CEO will need to organize the company in a more centralized manner.
- Competence–Even if a company aspires to operate as a more decentralized organization (Model B or C), it may have not developed competent business unit leaders.

CASE STUDY—THE EVOLUTION OF NEWMONT'S MANAGEMENT MODEL

William Boyce Thompson founded Newmont Mining Corporation in 1921. As a financial entrepreneur and promoter, in the early days Thompson built a mining financial holding company (Model A). His basic formula was to find potentially profitable mineral properties; form companies to finance, develop, and manage them; provide the sale, usually to the public, of new issues of stock in these companies; and build a liquid, income-producing portfolio of mining and petroleum securities to use in financing still greater mining enterprises.

The breadth and expanse of Newmont's investments grew, and the mineral financial model needed to be modified so that a core, albeit small, corporate staff provided strategy and oversight along with sourcing new opportunities and furnishing high-level technical consulting. This approach lies between Models B and C and was employed between the mid-1920s and through the early 1950s.

With Plato Malozemoff's accession to Newmont's presidency in 1954, a change in management structure became evident. The old horizontal divisions or geographic splits began to disappear. A vice president was put in charge of all production operations, another supervised all metallurgical developments, an engineering department was formed, and a worldwide exploration head was established. Before Malozemoff took over, each Newmont executive oversaw a particular area or division of the business, and there was virtually no staff on which to call for help. Malozemoff hired a small staff with expertise in geology, geophysics, mining, metallurgy, mechanical and electrical engineering, mineral economics, taxation, transportation, marketing, sales, and finance. Therefore, Newmont became more centralized. Although it was not a true command and control organization, it was more akin to Model C but with active core staff involvement.

After Malozemoff left Newmont in 1986, Gordon Parker took the reins. Under his leadership, the company generally became more centralized in its management approach while continuing to grow and expand its staff. However, in 1987 Newmont came under attack by T. Boone Pickens. As a result, it divested itself of a number of its assets, including Magma Copper and its interest in Peabody Coal Company. This resulted in Newmont narrowing its size and scope, with a focus on gold mining at its Carlin, Nevada, operations. It also acquired two large and influential shareholders, George Soros and James Goldsmith. These shareholders did force management to downsize corporate staff, but essentially a centralized management approach remained.

Starting in 1992, Newmont began expanding overseas, first with the development of the Yanacocha mine in Peru, followed by the Zarafshan joint venture in Uzbekistan and the Minahasa mine in Indonesia. Under Ron Cambre's leadership starting in 1994, Newmont acquired Santa Fe Pacific Gold in 1997, which bolstered the company's Nevada gold-producing assets. Newmont also developed the large copper/gold Batu Hijau mine in Indonesia. Over the last 6 years, the Yanacocha mine grew dramatically to become the largest gold operation in Latin America, producing nearly 2 million ounces per year.

Newmont's three core operating areas are significant businesses in their own right. Indonesian operations (Batu Hijau and Minahasa) generate revenues of nearly $700 million per year from copper and gold production. Nevada operations produced nearly 3.0 million ounces in 2000 and generated revenues of more than $800 million. Finally, Yanacocha continues to grow and generates revenues exceeding $500 million per year. The smaller Zarafshan produces about 400,000 ounces per year and generates slightly more than $100 million per year, but it remains a building block for further efforts in Central Asia. Given the size and diversity of Newmont's current operating portfolio, the company has naturally become more decentralized, with regional vice presidents overseeing their respective operations.

Newmont's corporate staff totals approximately 140, supporting a managed employment base of more than 8,000 people worldwide. Corporate functions include legal, accounting, finance and treasury, investor relations, public relations, corporate human resources, information systems, business development, internal audit, and administrative services. In addition, within corporate there is a small environmental services group, although most environmental efforts are managed at sites, and a small materials management group that strategically sources supplies through worldwide supplier alliances and forward purchase agreements.

Beyond the corporate functions, Newmont maintains a significant technical presence at the Plato Malozemoff Technical Facility south of Denver, Colorado. The technical groups include metallurgical services, exploration technology, exploration acquisitions, mining services, technical computing and engineering and construction. In total, about 80 people work in these groups.

It is appropriate that the technical facility is named after Plato Malozemoff, because he believed in a strong technical consulting and research and development presence. All these groups support existing operations and exploration efforts. They also assist in acquisition evaluations and in metallurgical and exploration technology groups where a small amount of applied research and development is conducted. Newmont prides itself on its technology development efforts, which have led to metallurgical breakthroughs and new geophysical exploration technology. Newmont holds numerous patents on aspects of processing refractory ores, which have created a competitive advantage and have led to utilization novel flotation, autoclaving, and biooxidation methods.

Although Newmont has become more decentralized from an operating and administrative perspective, management believes that a strong centralized technical group adds value. This group provides (1) a sustained repository of institutional knowledge, (2) expert support for operations without hiring consultants who take longer to climb the learning curve, (3) continuous development and learning opportunities for technical staff worldwide, (4) transfer of knowledge and best practices, and (5) better ability to undertake long-term corporate strategic research efforts that would not be tactically important to individual operating sites.

Newmont's management model continues to evolve. Recently, it was decided to have all site-focused pit/stope geologists and environmental staff report through their specific site general managers rather than through the corporate exploration and environmental departments, respectively. In addition, each of the core operating areas (Nevada, Peru, and Indonesia) have been working to strengthen their independent financial management, external affairs, and legal capabilities. Nevada does, however, obtain its legal support out of the Denver office and some support for government affairs out of the Washington, D.C., office.

Newmont has also been utilizing what it calls "Gold Medal Performance" (GMP) across the company as a cultural change and cost reduction program. GMP's primary focus is to drive out inefficiencies in Newmont's business. It has also allowed for better communications and the sharing of best practices between operating sites through global teams that focus on key areas such as load and haul, drill and blast, processing, mine maintenance, mill maintenance, and safety, among others.

In addition, Newmont has restructured its compensation systems to bestow bonuses more closely aligned with site and SBU performance, with a lower percentage related for corporate objectives. Denver-based staff will continue to be compensated relative to corporate objectives. In essence, Newmont operates under the Model C format with regard to SBUs but with a core consulting function.

CONCLUSION

Management models differ between mining companies. Most, as they grow, become less centralized and tend to move toward establishing either product-related or geographically focused SBUs (i.e., the typical business model.) However, one size does not fit all. Each individual company has its own issues combined with its historic culture, which has determined the appropriate model for it to follow. Newmont has evolved in regard to its management model. It has gradually become less centralized over the last few years, while retaining its heritage of maintaining a strong centralized technical support and consulting group.

. . . .

Political Metallurgy: "Sandbagging" Versus Prudent Caution, and Other Political Considerations

John O. Marsden*

ABSTRACT

The project development metallurgist is faced with an interesting dilemma: The use of overly conservative metallurgical projections and cost estimates can kill a project that has the potential to provide needed jobs and growth to a company's bottom line. On the other hand, if the metallurgist uses a highly aggressive approach by trimming capital and operating cost estimates to the bone and by applying highly optimistic metallurgical projections, there is a risk of creating a project that can never live up to expectations. The metallurgist must walk a tightrope between "sandbagging" and "overoptimism," and this tightrope is fraught with political intrigue and difficulty. This paper takes a somewhat lighthearted look at some of the political aspects associated with this issue, and other issues, and offers a few guidelines for walking the "political metallurgical" tightrope.

INTRODUCTION

In looking up the word "politics" in *Webster's Dictionary* (Tenth Edition), we find the phrases "the methods used or tactics in managing a government or state" and "intrigue or artful maneuvering within a group." The author proposes that "politics in metallurgy" (here termed "politico-metallurgy") is concerned with the methods used or tactics in managing metallurgical activities, which may involve intrigue or artful maneuvering within a mining or metallurgical group, or both.

Since the dawn of metallurgy sometime between 4000 and 3000 B.C., it is likely that metallurgists and their art have been plagued with political intrigue and subterfuge. It is easy to imagine some young Macedonian engineer, freshly graduated from the ficticious School of Mines in Thessaloniki with a masters degree in some cutting-edge field of gold metallurgy, being sent off by Alexander the Great to evaluate a gold prospect in a remote location in Asia (perhaps Uzbejkistan). Imagine their dilemma when they get to their destination and find gold, indeed lots of it, but much of it occurring in a frustratingly refractory form, at least refractory to the methods available to them at the time. Do they report that there is gold, but

* Vice President, Technology & Development, Phelps Dodge Mining Company

not much of it can be recovered economically (and risk exile to some other prospect even further afield)? Do they say that they need more time and more people to work on the problem? Should they speculate that they can solve the problem over time and report that this is a good project that just needs the money to develop it? Do they send out horseback riders to colleagues (Phi Delta Gamma alums?) in other reaches of the Macedonian empire in the hope that they may get back news of similar deposits and how to treat them 6 or 9 months later? Perhaps the gold is "unassayable," creating a range of additional problems for the unfortunate pioneer. Even worse, perhaps the owners of the deposit have a unique process to recover the gold that requires the addition of a magic ingredient that the owners will agree to supply at an outrageous price. How the plight of the young engineer must have hung in the balance and depended upon his ability to play out a game of metallurgical politics with his superiors and peers!

Moving back to the present day, perhaps not making such a huge leap as one might think, I have selected eight political issues facing the metallurgist active in today's mining industry. This is followed by a brief set of guidelines to help the metallurgist negotiate his or her way through the politico-metallurgical minefield.

ISSUE 1: "SANDBAGGING" OR PRUDENT CAUTION?

Probably the biggest political issue facing any metallurgist is how to provide a balanced approach to the metallurgical analysis and evaluation of a new project. On the one side, there may be a tendency to be overly optimistic for any number of reasons. On the other side, excessively pessimistic and conservative projections may be made, also for a variety of reasons. Metallurgical politics can play a significant role in how this approach is developed.

Let's first review the case for overoptimism. The prospect of future employment and career development for the metallurgist on a particular project may be one of the strongest reasons for overoptimism. Indeed, it is not only the project metallurgist's prospects that may be at stake, but also the livelihoods of many potential employees for the particular project. Project economics depend heavily on the metallurgical performance and cost projections. Metallurgical recoveries are directly proportional to revenues. With the exception of large, underground mining projects, the processing capital and operating costs typically account for between 40% and 60% of the total capital and operating costs. The more optimistic the projections, the better the discounted cash flow analysis results will appear.

Personal pride and professional credibility are other contributors to overoptimism–if the metallurgist cannot offer satisfactory metal recovery and costs to provide sufficient incentive and justification for the project to go into commercial production, he or she may be seen to have failed, losing credibility. This is particularly true when the expectations of senior management are high, perhaps sometimes unrealistically so, to the extent that the subordinate metallurgist feels compelled, and possibly pressured, to furnish the most optimistic metallurgical case to support the development of the project to an unreasonable extent.

Another reason may be lack of experience and skill: The metallurgist must take the results of test work and other relevant data and extrapolate these to generate estimates of metallurgical performance and costs at full commercial scale. The quality and accuracy of these estimates depends on the ability of the metallurgist to extrapolate and scale up the relevant data, and this ability in large part depends on the type, amount, and quality of practical experience and lessons the metallurgist has learned over time. Much of this type of experience is not, and in some cases cannot be, taught in a university or college class. One example of this is found in heap leaching of gold ores with cyanide. The most relevant metallurgical testing may include cyanide leaching in columns and pilot-scale heaps using representative samples of the material at the size that is expected to be treated in practice at commercial scale. Even though the tests may be conducted with absolute care and with exceptional precision and accuracy, the results cannot be applied directly to predict the gold extraction rate, ultimate gold recovery, and reagent consumption (hence costs). Scale-up factors based on experience must be applied to provide more accurate estimates of performance and costs at the commercial scale. The judgment as to what is representative, the effective recognition and identification of different ore types within a given deposit, the way in which each ore type is tested, and the manner in which the results are interpreted–all add further complexity to this exercise. This is discussed in more detail later. The real point to be made here is that it is often politically easier for the less experienced metallurgist to portray more optimistic metallurgical performance to senior management than to try to justify more conservative projections, even though these may more accurately reflect the expected performance at full commercial scale.

Let's turn now to the reasons for an excessively pessimistic or conservative approach to the projection of metallurgical performance and costs for a given project. As discussed above, the history and experience of the metallurgist responsible for this activity plays a key role in determining his or her capability in making metallurgical projections and applying scale-up factors. However, a metallurgist who has been "burned" on a previous project as a result of the inability of the commercial operation to meet the predicted performance targets, will be likely to be more conservative with his projections on future projects. The only thing worse than a failure is repeating the same mistake later. The expression "If we don't learn from history, we are destined to repeat it" rings particularly true here.

Even the metallurgist who has not been "burned" on a prior project usually wants to set up the commercial operation to be a success. Generally, he or she will approach the analysis and evaluation of metallurgical data with prudent caution. This means that the metallurgist must clearly understand the following:

- The different ore types and their abundance and location in the deposit
- The metallurgical characteristics of each ore type and variability of metallurgical performance for each
- The metallurgical performance for the ore types when processed together in the proportions predicted from the mine plan over the mine life (or at least over the length of the mine plan)
- Scale-up considerations for each of the metallurgical tests performed
- Process flowsheet options and their expected performance
- The predicted consumption of energy, reagents, wear steel, and other major supplies

The problem is when does prudent caution become "sandbagging"? A key issue in answering this question is that the application of scale-up factors to predict commercial performance from metallurgical test results for many processes is an art rather than a precise science. Two excellent examples of this follow.

Example A—Design of Semiautogenous Grinding Circuits

The results of metallurgical test work (e.g., pilot-scale semiautogenous [SAG] grinding tests, crushing and Bond Work Index tests, standardized abrasion tests, drop weight tests, and other SAG performance indicator tests) are used as the basis for grinding circuit design. This information is usually evaluated in the context of experience from other grinding circuits on similar ore types and third-party interpretation and modeling to result in recommendations to the project metallurgist for circuit configuration and design. However, the final mill dimensions and motor size selection are imprecise, and the rated throughput of a given mill configuration and power must be based on the best judgment of the metallurgist involved. A conservative approach may result in the mill or drive motor, or both, being oversized for the required throughput. Conversely, of course, an overly aggressive design may have the reverse effect, resulting in the facilities being underdesigned for the proposed purpose.

Example B—Design of Heap and Dump Leaching Operations for Gold, Silver, Copper, and Other Metals

As mentioned earlier, the design of heap and stockpile (dump) leaching operations is generally based on column leach testing data and sometimes pilot-scale heap/dump leach tests, combined with detailed mineralogical information and detailed metallurgical data for a large number of samples throughout the deposit. Building on this example further, scale-up factors need to be used to allow for the following:

- Delays in solution breakthrough out of the bottom of the heap for multiple lift heap operations
- Increasing holdup or inventory of solution (and dissolved metal) in the heap over time
- Side slope effects because of less efficient leaching of these areas, if applicable
- Inefficiencies of solution application on the heap (dry spots resulting from plugging of drip emitters or uneven application of sprays)
- Inefficiencies of solution contact with the ore as the solution flows through the heap (as a result of channeling, migration of fines/clay material resulting in regions of high/low permeability)
- Climatic conditions, such as freezing, snow, and excessive rainfall events, among others

When all these issues are taken into account, the metal recovery rate and the overall metal recovery achieved in the commercial operation will be different from those achieved in column leach tests. However, the scale-up factor for each of these depends on the ore type(s), the design and configuration of the leaching system, the experience obtained from other similar deposits and processing plants, and many other factors. This requires judgment on the part of the metallurgist in addition to the application of sound science.

One other factor that should be mentioned in this discussion of reasons for an overly pessimistic or conservative approach is that such projections are easier to make and justify when the project in question appears to have rather favorable economics. Stated differently, a project with attractive economics can sustain more conservatism in the metallurgical assumptions than a project with borderline economics.

ISSUE 2: CONTINUITY BETWEEN PROJECT DEVELOPMENT AND OPERATIONS STAFF

The need for effective continuity between the project development metallurgist and the metallurgist responsible for startup and operation of the facility is critical. Although this doesn't need to be a political issue, it often is. From a political point of view, the best case is when the project metallurgist responsible for analyzing and evaluating the test work results is to be maintained in a line-management function throughout the startup and operation of the commercial facility. This requires that sensible goals and compensation/bonus schemes be provided to the metallurgists to reward them for good performance, rather than rewarding them for conservatism or optimism during the development phases of a project. This sounds obvious, but a goal to "achieve a favorable outcome from a project feasibility study by a certain deadline" does not always result in a successful project. Continuity may also be effectively augmented when an independent third-party metallurgical consultant is employed to assist with, and complement, the analysis/evaluation by the project metallurgist and to help transition the project into operation. On the other hand, an adverse political situation may occur when the project metallurgist performs his or her duties knowing that he or she will have little or no involvement in the commercial application and that compensation received is based on factors unrelated to the accuracy of their predictions. Worse still, compensation might be based on how favorable the metallurgical performance and cost projections are at an early stage in project development. This "worst case" scenario would be further compounded by the lack of any independent third-party metallurgical review. One can only pity the poor operators in such cases, who must try to salvage what they can from the original unrealistic projections and with virtually no opportunity for any project upside with which to meet their senior management's high expectations.

ISSUE 3: HANDLING THE "NOT INVENTED HERE" SYNDROME

The reason that many good metallurgical inventions never come to commercial fruition is the "not invented here" syndrome. All metallurgists need to develop effective techniques to communicate innovative ideas to others who may be less receptive than they expected to their latest "brilliant" development. The best metallurgical consultants understand that metallurgists working for mining companies and corporations, whether in operations or in a project development role, often do not have the time or the resources to look at a problem from the best problem-solving perspective. There may be a lack of time to devote to the problem, a lack of sufficiently diverse experience (whereas the consultant may have seen the problem before), an inability to look at the 'big picture' and put the problem in true perspective, and a lack of qualified staff. Effective consultants are able to make the company metallurgist feel valued and respected by communicating their understanding for the metallurgists' situation and by getting buy-in from all sides. The key to this is to make the company metallurgist feel that he or she has participated in supplying part or all of the solution. The best political metallurgists have worked out ways and means to change the "not invented here" syndrome to "we invented this here, with some help from a great consultant/metallurgist."

ISSUE 4: LEVERAGING METALLURGICAL "SYNERGIES" (THAT CAN'T BE ACHIEVED IN PRACTICE)

Sometimes, the leveraging of metallurgical "synergies" is used as the reason for predicting that actual recoveries at the commercial scale will be significantly higher than those predicted by test work. For example, let's consider an ore deposit that has highly variable metallurgical response for a number of different ore types in different locations within that deposit. This might be due to variation in mineralogical associations or variability in clay content, either of which may greatly affect the metallurgical performance. The metallurgical response for each ore type might be well defined by detailed metallurgical test work and also by testing various blends. The metallurgical performance achieved for the optimal blend is projected to be the performance achievable in practice. The performance of the blend represents the "synergy" between the treatment of the different ore types when mixed together. The blend gives much better performance than the poorly performing materials, and sometimes the blend has an expected performance that is projected to be equivalent to the "best" ore type treated on its own.

Several potential problems are associated with this assumption. First, the mine plan must be developed specifically to ensure that the optimal, or close-to-optimal blend, can be delivered on a routine basis. Secondly, the mine must be able to operate cost effectively to consistently supply the optimal ore blend to the processing facilities. This is becoming increasingly difficult to achieve in operations where the size of mining equipment has increased significantly in recent years, resulting in more difficulty with selective mining and limitations in the number of active ore faces being mined at any given time. A number of copper and gold operations operate with one shovel in ore at one time. This is a political issue, because the issues associated with blending requirements and the capabilities in the mine to meet a projected metallurgical performance must be addressed by reserve and resource estimation, mine planning and design, ore control, mining operations, metallurgical design, and metallurgical operations staff working together. All this must be done while managing budgets and cost centers that may have conflicting goals and objectives. For example, minimizing mining costs is not synonymous with optimizing ore blending to the processing plant or with maximizing the utilization of a primary crusher.

ISSUE 5: HANDLING THE MINE-TO-MILL DISCREPANCY

The mine-to-mill discrepancy is probably the most important political issue facing the metallurgist working in operations because it represents an accountability "no-man's-land" in which an unsuspecting metallurgical engineer may find himself caught. Ore reserves estimation and mine planning can generally be done with good precision and accuracy if adequate data are gathered for a given deposit. Also, estimates of tonnage and grade of material to be mined can be projected with reasonable certainty, if accurate assumptions are made about mineability and dilution. At the other end of the operation, the weights and assays of the final salable metal-containing products are measured to close the metallurgical balance for the facility. For the processing steps leading up to the final products, estimates of performance and efficiency can generally be determined with some level of accuracy and precision using established sampling, analytical, and material balance calculation procedures.

The real problem, however, lies in the difficulty of accurately and representatively sampling bulk, as-mined (run-of-mine) materials before the metallurgical processing facilities. The difference between the tonnage and grade of material delivered to the processing facilities from the mine (estimated using truck fill factors or truck weights, truck counts, and blast hole assay data) and what is reported by the processing facilities (i.e., using tonnage measured by conveyor weightometers and grade by automatic sample cutters on conveyor belts or across slurry streams) is known as the mine-to-mill discrepancy. For a well-run facility, the discrepancy may vary between 0% and 5% and generally averages between 2% and 3%. This discrepancy almost always indicates less metal value delivered to the processing facilities than estimated by the mine (i.e., a negative discrepancy). A mine-to-mill discrepancy exceeding 5% negative indicates a serious problem with the mine-to-mill metallurgical balance that needs to be addressed. In most cases the reasons for the discrepancy are:

- Higher than expected dilution in the mine resulting from a less effective mining selectivity at the face than expected
- Blast hole sampling bias
- Inaccurate truck factors
- Incorrect apparent density assumptions for the ore
- Conveyor weightometers in poor condition, uncalibrated, or dirty
- Incorrect estimation of moisture of ore delivered to the processing facilities
- Unrepresentative sampling in the processing facilities
- Combinations of the above factors (most common)

The only way to address this political issue is to rigorously apply established and generally accepted methods to result in as accurate projections of tonnage and grade as possible at all critical steps in the operation and to communicate this effectively between mine, processing personnel, and senior management. Again, this is difficult to do when the goals and objectives of the different departments within a given organization are not aligned.

ISSUE 6: POLITICAL ENGINEERING

It is well known that the engineering design, procurement, construction, and construction-management side of the mining business is highly competitive. As a result, there is always the possibility that a contractor may inadvertently, misguidedly, or perhaps even deliberately, low bid a project to get the job, and then grossly overspend, resulting in a "blown" budget for the project. The only ways to monitor and control adverse political engineering activities and prevent this type of political catastrophe are to:

- Develop a well-defined scope of work for the project.
- Work diligently to minimize variances from the scope of work and require project staff to develop variances (both positive and negative cost impact) such that there is a zero-net-sum cost impact to the project.
- Have experienced staff dedicated to monitoring the engineering, procurement, construction, and construction-management contractor activities throughout a project.
- Have effective cost controls in place that are rigorously monitored and acted on throughout the project.

ISSUE 7: CHAMPIONING PROJECTS

Every project needs a project champion–someone who promotes and drives a project to get to a final conclusion in an effective and expeditious manner. This has to be someone who believes intuitively or instinctively that the project in question has merit and that the project will succeed. In many cases, this can be a politically charged issue, particularly if the project is not universally supported within the organization. However, occasionally someone may develop an unreasonable, unjustified, and completely unexplainable emotional desire to see an uneconomic project go into production. There is no simple or easy answer to this issue, which is often hard to recognize and deal with. One way to address this is to periodically do reality checks with other metallurgists familiar with similar project situations.

ISSUE 8: CRACKPOT MANAGEMENT

The mining business is a truly a small world. The metallurgical side of the business consists of a well-connected group of professionals working diligently to improve processes and operations. However, occasionally an individual or a small group of people approaches the mining industry with a novel, innovative, and potentially performance-enhancing product or process. The problem is that for every genuine idea that can be turned into a truly groundbreaking and worthwhile development, there are 10 or perhaps 50 ideas that do not turn out to be as performance-enhancing as advertised. The metallurgist needs to be able to evaluate and make a decision on these new products or processes rapidly to avoid getting bogged down in non-value-adding activities. There are a number of ways to get to the decision-making point, including the following:

- Perform some preliminary (cheap and quick) tests on the product or process.
- Equate the product or process to prior experience with a similar product(s) or process(es).
- Get help from a reliable, trusted metallurgical consultant who has experience with the process or product itself, or with similar process(es) or product(s).
- Form an opinion of the character and motives of the person promoting the product or process and weigh this heavily in your decision making.

The important issues here are that it is easy to waste a lot of time and effort in pursing blind alleys and red herrings and that even the odd missed opportunity may be less important than the ability to efficiently weed out the worthless ones.

GUIDELINES TO HELP ADDRESS POLITICAL METALLURGY ISSUES

Finally, this paper closes with a few guidelines that the author hopes will help the uninitiated metallurgist tiptoe his or her way through the minefield of political metallurgy:

- Make sure you have all the relevant and necessary data before making projections of metallurgical performance and costs. If you have to make preliminary projections, make sure that they are accompanied by effective qualifying statements.
- If you don't have the data you need, say so, and make clear recommendations and plans for the work that needs to be done.
- Seek well-qualified and experienced independent metallurgical consultants or contractors to give you an independent third-party review and opinion. Make sure the consultant or contractor doesn't have an agenda or an axe to grind in the area for which you are asking them to consult. If you do not have the authority to solicit such third-party assistance, offer a strong recommendation to your manager or superior that such third-party assistance is required, clearly stating the reasons for your recommendation.
- Always write down your recommendations and conclusions when making projections of metallurgical performance and costs. The recommendations should be clear, concise, and brief, and should be easily readable and understandable by people at all levels in the company or corporation.
- Never burn bridges. The mining industry is relatively small and everyone on the metallurgical side of the business knows everyone else.
- Never tell your boss that his or her good friend's latest metallurgical innovation is a bad idea. Rather, after careful consideration, you will probably determine that it is a very interesting concept but one that is not economically attractive at your operation(s) under the prevailing economic conditions.
- Finally, at risk of committing an oxymoron to paper, always be truthful, honest, and open in your political metallurgy dealings.

Do Managers Use Consultants Because They Are Smarter?

Paul Chamberlin*

ABSTRACT

Management has assorted roles to play, many of which are listed, that may call for consultant input. Consultants also play various roles, and some of these are discussed as well. The relationship between management and the consultant depends mainly on management's preparation for the assignment and on their monitoring during the assignment. Consultant costs are discussed briefly and suggestions for keeping consultant costs under control are given. In addition, some qualities of a good consultant are covered.

INTRODUCTION

A consultant is an expert who is called on for professional or technical work or opinions. Nothing wrong with that. Seems reasonable. All of us are supposed to gather relevant opinions and get the job done. That may include help from a consultant. Managers are not supposed to have the "not-invented-here" mind-set. But, as a manager, do I really want to hire a consultant and run the risk of upsetting my staff?

The title of this paper implies that management may think consultants are smarter than their own in-house personnel and experts. Not so! Consultants may have more experience or expertise than in-house people, but they aren't smarter. Any consultant who thinks he or she is takes a big step toward failure. So why would an astute manager want to hire a consultant?

This paper explores the relationships between management of a company and consultants. It will not be an exposé of inept consulting or an operating manual for those who want to join the consulting ranks. It will explain why management hires consultants and how management can make that relationship work better.

ROLES OF MANAGEMENT AND CONSULTANTS

Management Roles That Call for a Consultant

Management is a multifaceted task that requires looking inward to the day-to-day operations and resources of the company as well as looking outward to all the external influences on the

* Chamberlin & Associates

company. The range of inputs is very wide and complex, and the pressures for performance are great. In-house resources may not be sufficient. Uncertainties may call for more input, and there is always the desire to minimize risk. For a host of reasons, management may want to call on consultants.

- Management wants to keep annual costs low, especially when certain activities occur infrequently. In spite of the consultant's higher hourly fee, hiring a part-time consultant to perform these infrequent activities is cheaper than hiring a full-time employee to do them. This can be a sensitive issue if management is terminating full-time employees and replacing them with consultants on a part-time basis.
- Management wants to reduce the risk of not having the best answer, not meeting budgets, and not meeting production. Consultants may alleviate those risks.
- Management has initiated so many projects with short deadlines that there is insufficient in-house labor to complete them. The agenda has become so crowded that outside help is needed.
- Two management groups within a company may have different opinions about a matter and an impartial opinion is needed to minimize the growing conflict. This calls for a consultant.
- Management may want an opinion substantiated or critiqued by an independent outsider.
- Management may want to remain anonymous in a matter. A consultant can conduct an investigation without revealing the company's intent.
- Management has encountered a project that calls for specific skills that in-house people do not possess.
- Management wants an independent appraisal of their current operations to improve them and to avoid problems.
- Management wants to try something, but it may be politically risky. Hire a scapegoat—get a consultant to propose the idea and if it doesn't fly, management is off the hook.
- Some smart managers realize that the older they get, the less certain they are about some matters in this complex minerals business. This is a sign of wisdom. Therefore, they may hire a consultant to play "devil's advocate." Let the consultant point out the flaws in proposals and projects or confirm their soundness.
- Management should always expect that hiring a consultant will pay dividends, perhaps ten times the consultant's costs. Consultants' costs include not only the fee and expenses but also the cost of the time company employees spend with the consultant, the cost of tests suggested by the consultant, and the cost of changes that may be warranted.

ROLES OF A CONSULTANT

- Consultants are an unbiased resource who render opinions and perform work solely in the interest of the client.
- Consultants offer new ideas and perspectives.
- Consultants provide information.
- Consultants stimulate thinking within the client company by asking questions and by carrying thoughts from one department to another.
- Consultants improve communications among plant employees. Excellent ideas often come up in conversations between the consultant and client employees. Neither knows in advance where these gems may lie; often they come from client personnel. Follow-up on these ideas leads to better interdepartmental communications.

- Consultants may act as a sounding board for new ideas. They may hear confessions about poor operating practices because the employees don't expect to see the consultant again.
- Consultants who are senior and experienced usually provide a substantial, positive impact on long-range profits by guiding the strategy or preliminary design of a project.
- Consultants often tell management the reality of a situation even if management doesn't want to hear it. This may serve as a wakeup call to management.
- Some consultants make good expert witnesses. This type of assignment requires being an expert, a teacher, and a good communicator. Teaching the attorneys and jurors is critical; the more persuasive the consultant is, the better. A good consultant, retained early in a case, can counsel management on whether or not an extra expert is needed to testify. Management may decide to hire one expert to serve as a consultant and a second expert to testify, insulating the two so that the consulting expert's work is not subject to discovery. Early involvement of a consultant can help prepare a complaint, spot defenses to a lawsuit, allow better preparation for testimony, help formulate a discovery plan, promote the review of documents, and be of help during the deposition of opposing experts.

RELATIONSHIP BETWEEN MANAGEMENT AND CONSULTANT

Management and the consultant must have a relationship of mutual respect with very open communications. Before hiring the consultant, management must clearly define the consultant's scope of work. This includes agreeing on a step-by-step process to achieve management's goals. To prepare the scope and the sequencing of activities, management may need to interview consultants to determine how they intend to achieve the desired results. Management must insist that the consultant's work achieve closure; some consultants may tend to leave a project open-ended so that it leads to the next project. The obligations of both consultant and management need to be voiced or put into writing. Some of the subjects to be addressed are schedules, dates, reporting relationships, and invoicing procedures.

Hostility between a client's employees and a consultant can be avoided. It is up to management to inform employees that a consultant will be retained and why. It is desirable to have an introductory meeting at which management introduces the consultant to key employees, explains objectives, and specifies procedures. It is up to the consultant to work through the designated employees and to give them credit for their time and contributions.

Management should examine its motives. In other words, don't hire a consultant to rubber-stamp changes that you intend to make. Don't use the consultant to sell your agenda to the Board of Directors. Allow the consultant to draw his or her own conclusions.

Management may have concerns about hiring a consultant. They may fear that the consultant will turn out to be incompetent or that their own in-house people will be incapable of properly directing the consultant. There is always the worry that the consultant's fee is too high or that the consultant is too busy to expend the time required to do a good job in the time allotted. Human nature being what it is, in-house personnel often feel that they can capably do whatever their bosses want done; they may think that hiring a consultant is an expression of their failures or limitations and an admission that something is wrong. Management may fear the loss of proprietary or sensitive information through the consultant. All these fears, worries, and concerns should be addressed verbally before the consultant is retained for the assignment. Putting some of them in writing may also help.

COSTS

Consultant Versus Employee

The average consultant in the minerals industry is underpaid compared to his "employed" peers in industry. Several underlying assumptions for this opinion are explained in detail in a paper prepared by Allan Juhas titled "An Income Nomograph for Geologists," which is being presented at one of the sessions during this meeting. Some of the assumptions are:

- Consultants of various ages, educational levels, and position levels should take home about the same each year as their regularly employed peers.
- Both the consultant and the employed peer are about the same age and have the same level of experience.
- The consultant accomplishes in 150 billable days per year what the employee accomplishes in 250 8-hour days per year. The "useful work time" from employees is in the range of 60% of their total work hours, whereas consultants can honestly invoice only for the useful work time that they provide. This can be a gray area, but consultants know that hours worked and billable hours are seldom even close to being the same. It is a truism that most consultants are more committed and obedient to their clients than employees are to employers.
- The comparable pay for consultants, as calculated by Allan Juhas, includes only direct pay and some of the fringe benefits. It does not include many other items that employees get from their employer such as professional training, education, convention costs, memberships, time for outside meetings, time for professional volunteerism, time for technical research and keeping up with developments in business and technology, proposals, marketing, insurance for disability or long-term care or liability, sick days, the costs of office space and office overheads, and capital equipment costs for the office and field.

An example from Juhas's nomograph shows that a consultant working 150 days per year should be paid $1,200 per day if he interacts with and reports to the vice president of a mid-sized company who draws an annual salary of $145,000 per year plus fringes. The consulting rate should actually be higher to cover the items listed in the previous paragraph that the employee gets in addition to his pay and fringes. In addition, consultants may have expertise the employer cannot deploy internally, which should be worth a premium. Further, the consultant is typically "on standby" and can be called on for immediate deployment by the client. This availability and the resultant peaks and valleys of consultant demand should also be worth a premium.

Companies are fortunate these days because there is such a large pool of consultants from which to select help. Average consulting costs are low because of the oversupply of consultants caused by corporate layoffs and acquisitions. Many of these consultants will move out of the minerals industry. When this occurs and when the next boom period occurs in the minerals industry, companies will find themselves paying more for consultants than they do today.

HOW MANAGEMENT CAN KEEP CONSULTANT COSTS REASONABLE

- Define what you expect of the consultant. Communicate, communicate, communicate! Do you want a troubleshooter, a trainer, a mentor, a hard worker with expertise, a temporary employee, a resident sage, a father confessor, a witness, or a writer?
- Insist on adequate preplanning of the consultant's assignment. If your consultant hasn't told you what he or she will do and why, how do you know that his or her activities will be in your best interest?
- Invest in people, not firms.

- Invest in demonstrated expertise. Check your communications network for consultant reputations.
- Competitive pricing is generally not the most cost-efficient way of investing in consultants. The most expensive consultant is often the one with the wrong answer.
- Routinely manage your consultants. Keep in touch with them. The squeaky wheel gets the grease.
- Require your consultant to track schedules and costs.
- Always get a final report. Require clear, decisive conclusions even though consultants may not like to make them.
- Use a consultant who is busy.
- Use a consultant who wants to use your people where possible.

FALLACIES ABOUT CONSULTANTS

Fallacy Number 1

Consultants are mind readers. Nothing is further from the truth. Yet management often assumes that consultants know what is expected of them. It is the wise manager who explains in slow detail the situation that the consultant is entering. It takes more than 5 minutes to bring the consultant up to speed and to negotiate a clear agreement about what is expected of the consultant.

Fallacy Number 2

Consultants are rich. Some are, especially in the legal and entertainment fields. The minerals industry, however, is not the place for most consultants to seek their fortune. Most consultants simply sell their time and only a limited number of hours per year are for sale. Fortunes are not made selling time. Assuming that the average consultant can find payable work for 150 days per year and charge $100 per hour, the maximum annual income may be $120,000 before expenses.

Fallacy Number 3

"I'll be a consultant when I retire." Sure you will ... if you wish to join all the others who are competing with you, if you have the required contacts and expertise, if you still have the drive to work very hard, if you are willing to work largely by yourself rather than within a corporate group, and if you are willing to work without any recognition. Consulting after retirement is usually akin to being a "gentleman farmer."

Fallacy Number 4

Consulting businesses are started by design. Maybe a few are. But the vast majority are accidents, especially today. For example, there's a company consolidation, you're out on the street, and you become a consultant. The company is downsized, you're out on the street, and you become a consultant. You can't find full-time employment, you take on short-term assignments, and you become a consultant.

Fallacy Number 5

Consultants just take the money and run. The resulting bad reputation will kill that consultant's business. He or she won't be around very long. To a great extent, the only thing that consultants have to sell is their reputation.

QUALITIES OF A GOOD CONSULTANT

You've heard it often: "Anyone can be a consultant when he is more than 20 miles from home." This has been quoted for a long time. Even 2,000 years ago it was written in Luke 4 that "... no prophet is believed in his own country." Pundits have been paraphrasing that sentiment ever since. Consultants are often held in contempt by their employed peers. "You can't do it yourself so you are going to be a consultant, huh?" "If you can't cut the mustard, you go into consulting." "Consultants talk a good line but the capable people actually work for a company and do it."

Are these sentiments justified or true? We all like to give advice and opinions, but does that make us good consultants? A list of good consultant qualities follows.

- Be honest. Do a day's work for a day's pay. Say what you believe rather than what may be politically expedient. Keep your word. If you say you will do it, don't back out. Be on schedule. Have the courage to tell clients the real truth, even if it hurts. Don't pussyfoot around with weasel words. Some clients need to be hit between the eyes with the truth.
- Do good work. Investigate all questions. Insist on getting the facts and checking their accuracy. Go the extra mile.
- Have a good sense of relevance; cut through the haze and concentrate only on relevant issues.
- Cultivate the ability to see all sides of a problem, especially if you are a more senior consultant. Have the capacity to take the long view, to anticipate remote and collateral consequences.
- Have the ability to write well and to speak with clarity and persuasiveness.
- Be well organized.
- Have the discipline to persist at a project in spite of distractions, especially those at the home office. Enjoy work. Be goal oriented.
- Be good at marketing, even if you don't enjoy it. There is no factor more vital to consulting success. Sell only your expertise. In the desire to make a living, consultants may tend to claim expertise in fields that they have experienced only peripherally. It is better to hire another consultant with the requisite expertise. Management should be wary of consultants who claim to do everything. They shouldn't hire the know-it-all.
- Respect your competition. You may want to work with them someday. Or, as your mother probably told you, "If you can't say something nice about someone, don't say anything at all." Complaining consultants don't get much respect. Watch your tongue.
- Know what you're getting into. Not everyone has the aptitude or personality to be a successful consultant–although that is hard to prove because of the number of them around today. One successful but unhappy consultant returned to research and development because he got lonely. Consulting is not a group activity for most consultants. They work alone, often from their homes. There may be little interaction with other people for long periods of time. It certainly is not the same as working in a company environment, chatting at the coffeepot and having group meetings.
- Accept being in the background. You may help your client make or save millions of dollars but you will get none of the credit.
- Be a good listener. Companies often communicate poorly between departments. The consultant is often the carrier of good ideas that already exist at the company but simply haven't been carried to the policy maker or implementer.

- Know when to fold and walk away. Sometimes management will just be bull-headed and insist on doing the wrong thing. After your repeated recommendations for a different course of action have been rejected, it is best to just excuse yourself from the assignment.
- Be a problem solver and teacher. Successful consultants are good at both of these tasks.
- Be able to live with uncertainty. A constant paycheck is not assured; neither is a constant supply of paying work.
- Maintain confidentiality about the client's work. Keep your mouth shut.

THE NEED FOR CONSULTANTS

Consulting has been with us since human gatherings began. It started when the boss caveman saw Fred Flintstone make a sharp stick and catch a fish with it. He asked, "How did you do that?" and consulting was born.

The consulting business is here to stay. The increasing complexity of ventures, both technically and sociopolitically, will make it difficult for most companies to have full-time employees covering all aspects of a venture. Only the largest companies will have this capability and even they, during periods of work overload, will call on consultants for emergency help.

CONCLUSIONS

Management has many reasons to hire consultants, and the reasons vary widely. It's not because consultants are smarter, but because they can assist management in many ways. Not just any consultant will do–it is management's task to find the appropriate consultant and to be sure that the relationships among management, company employees, and the consultant are effective. And finally, most consultants in the minerals industry are not charging excessive fees.

REFERENCE

Juhas, A. 2001. An income nomograph for geologists. Unpublished. To be presented at the 2001 SME Annual Meeting, Denver, Colo.

Do Engineers Make Good Managers?

Timothy D. Arnold, P.E.* **and James R. Arnold, P.E.†**

ABSTRACT

The mathematical rigor of the collegiate engineering experience prepares an engineer well for the technical aspects of his or her career. However, as that engineer progresses in the industry, how well does he or she do as a manager? The exacting, detailed, and complete problem-solving process that engineers are trained to use sometimes conflicts with the nimble, rapid decision-making process required in the dynamic world of business. This paper discusses the topic of "engineers as managers." It will investigate the personality traits of engineers relative to the traits of successful managers, and provide specific details about how well engineers have done in the field of management.

INTRODUCTION

Modern thinking in management and personnel theory is to compartmentalize people by personalities, attributes, leadership skills, backgrounds, etc. Everyone can fit in a "box," and there seems to be a lucrative market in creating boxes and promoting grandiose generalizations about those within the boxes. One of the popular methods of compartmentalizing people is by personality types, and one of the most recommend of these is the David Keirsey method. According to Keirsey, the coauthor of *Please Understand Me: Character and Temperament Types*, (Keirsey and Bates, 1984) personalities are generally grouped into Artisans, Guardians, Idealists, and Rationals. Keirsey and Bates use a matrix to further isolate 16 specific personality subgroupings. In studying the suggested careers that correspond with the subgroupings outlined in the book, 3 of these 16 subgroupings are considered ideal for engineers. Following is a partial list of famous persons who fall into these 3 "engineering friendly" subgroupings:

Franklin Roosevelt	Woodrow Wilson	Harry Truman
Theodore Roosevelt	Colin Powell	Brigham Young
Winston Churchill	Bernard Montgomery	Douglas MacArthur
John Paul Getty	John D. Rockefeller	Peter the Great
Charles Lindbergh	Dwight D. Eisenhower	Margaret Thatcher
Napoleon Bonaparte	Bill Gates	Madonna

* General Manager, Hecla–Lucky Friday Unit

† President and Chief Executive Officer, Knight Piésold and Co.

Although few of these people are or were engineers, they are or were undoubtedly successful as mangers of their careers and they have personality traits that some experts believe would be well suited for a career in engineering. In short, it seems that the type of personality that would make a good engineer is also the type of personality belonging to a very auspicious group of leaders and managers. At least from a personality perspective, the sets of good engineers and good managers are not mutually exclusive.

Another test of an engineer's ability to be a manager is trends in the industry. "Technology is becoming increasingly important to companies in all sectors in the economy," stated Alan G. Merten, Dean of Cornell University's Johnston School of Management. "In today's world, no industry is isolated from technology" (Payne et al. 1995).

In a 1995 study of 464 chief executive officers (CEOs) of the largest public industrial and service companies in the United States, 24% of all CEOs had technical or scientific backgrounds, as compared to 21% in 1985. If you were to add those with operations/manufacturing backgrounds to this group, many of whom have engineering backgrounds, the percentages would increase from 39% to 44%. The feeling is that as the economy becomes more knowledge-based, senior managers need technology backgrounds to understand business opportunities, limitations, and the implications of technology decisions. "Professionals who combine scientific or other technical knowledge with business training may indeed become the so-called 'gold collar' workers–a new generation of business elite with several distinct areas of expertise," Merten says (Payne et al. 1995).

So if the engineering personality is suitable for management and the technical background is not only complementary, but also increasingly important in management, what's left? What else does the engineer need to become a solid manager?

Consider the following hypothetical situation. You come to work in the morning as a unit manager and three people are waiting to see you. One is the operations foreman who tells you that the maintenance crews are upset because the microwave oven in the lunchroom has been broken for 3 weeks and no one has replaced it yet. They are talking about a work slowdown because management doesn't give a damn about the people. Your marketing manager gives you the results of a campaign you launched last spring entitled, "Widgets; You Really Need Them," and tells you that sales have dropped 30%. Your investor relations manager tells you that she is sick to death of working with that catty purchasing supervisor, and if you don't fire him today she will personally rip his tongue out and feed it to her pet piranha. Somehow the familiar green graph paper doesn't seem helpful today. Successful managers have learned to deal with issues far outside their area of expertise.

As a matter of fact, there is no degree that you can get that will give you clear guidance in the situation above. Let's look at two highly successful CEOs from different backgrounds.

JACK WELCH—GENERAL ELECTRIC (GE)

Certainly, Jack Welch is an icon among CEOs. He is "widely acknowledged as the leading master of corporate change in our time"(*Fortune* 1993). Welch comes from an engineering background. If one were to read the volumes written about him, you would think that he was to GE the "savior resurrected." Quite the contrary, Welch's immediate predecessor, Reginald Jones, retired as "the most admired business leader in America" (Slater 1993). In financial terms, such as profit growth, return on equity, return on sales, and return on assets, GE performed as well under Jones' 8-year tenure as during Welch's first 8 years (Collins and Porras 1994). Jack Welch comes from a long line of managerial excellence and he had an opportunity to learn from the best throughout his career. He changed GE, and so did his predecessors. His goals of being Number 1 or Number 2 in every market they served, having the strengths of a big company combined with the leanness and agility of a small company, all the while not compromising integrity, served him well. However, much of his training came after his years at the University of Massachusetts.

On several occasions, Jack Welch has tried to explain his methods of success. Here are his 10 "secrets" as outlined by Crainer and Dearlove (1999).

- **Invest in People.** People matter. Talking to them and meeting them takes up a serious chunk of Jack Welch's time. So, too, does developing people for the future. He talks, pontificates, cajoles, and educates. But, most of all, he connects.
- **Dominate Your Market ... or Get Out.** The choice is simple and is repeatedly laid out by Jack Welch. He has no time for companies that are fourth or fifth in their market. He wants to be first or a close second. Gain market leadership, take the market by the scruff of the neck, and lead it forward. If you can't get to the front, sell the business and look elsewhere.
- **Never Sit Still.** Welch is restless. Despite heart surgery, he never stops. He has inculcated GE with the same sort of restless energy. The company won't stay still or rest on its laurels. GE changes and then changes again. It is always nearer its goals by never staying in one place.
- **Think Service.** Pre-Welch, GE was a manufacturer, a good old-fashioned champion of the smokestacks. Welch introduced it to service. Now GE is a service company that also manufactures. It is a finance company and an information company as well as a maker of appliances. Quality and service link its activities.
- **Forget the Past, Love the Future.** For a company with such a great history, GE under Welch has become preoccupied with the future. It embraces the new–whether it be information technology (IT) or the Internet. Welch envisions the future. He speaks enthusiastically about the future, and GE creates the future.
- **Learn and Lead.** The new model leader is not a corporate dictator. The leader is committed to learning, deciding, and moving forward. Wrong decisions present their own opportunities. Learning from failure is more important than wallowing in success.
- **No Bull.** Jack Welch communicates. He is straight. Whether he is talking to workers in a GE factory, managers in a training program, or industry analysts, he speaks with passionate clarity. He tells it as it is.
- **Kill Bureaucracy.** Dismayed by the time wasting of bureaucracy and hierarchy, Jack Welch nearly left GE after his first year. He was talked into staying, but the bugbear remained. Since taking over at the top, Welch has eradicated bureaucracy with a vengeance.
- **Stick Around.** The corporate person is supposedly dead, but Welch has done all right sticking with a single employer.
- **Manage the Corner Store.** Welch manages GE as if it were a corner store. The same things matter to all businesses: quality, service, people, cash flow, and keeping abreast of what sells and what does well. The fact that you are selling nuclear power plants and not candy bars is immaterial.

HERB KELLEHER—SOUTHWEST AIRLINES

When a Northwest Airlines ad once used selective statistics to claim it was Number 1 in customer satisfaction, Herb Kelleher responded with his own ad: "Liar, liar. Pants on fire." To deliver his annual message to employees in 1996, the company's twenty-fifth anniversary, Kelleher was escorted to the stage in a straitjacket to celebrate the day's theme: "Still Nuts After All These Years." Kelleher, the Harley-Davidson riding, Elvis-impersonating, chain-smoking chairman, president, and CEO of Southwest Airlines, greets employees with a signature bear hug, and is as likely to be found tossing luggage with the baggage handlers or having a beer with an off-hours flight crew as he is to be in his office. It's hard to argue with the results: His $3.8 billion, Dallas-based airline is the only major U.S. carrier to have turned a profit every year since 1973 (Melymuka 2000).

"It's better to be Irish than smart," Kelleher is fond of saying. In the worst period in industry history, as other airlines lost a combined $12.5 billion between 1990 and 1994, Southwest earned $521 million. "Herb is truly one of the giants of the industry," says Ron Woodard, president of The Boeing Co.'s Commercial Airline Group (Enbysk 1995). Often referred to as the best CEO in American industry, Kelleher is a New York University Law graduate.

Although Kelleher is not an engineer, there are similarities between his ideas for success and those of Jack Welch (quotes taken from Enbysk 1995):

- **In Choosing Your Employees, Attitude Is More Important Than Education or Expertise.** "A good attitude is most important. It's something you can't change, but you can teach people to do whatever job you want them to do." In choosing managers, he seeks out leaders rather than bureaucrats or "suckups."
- **Pay Attention to Your Employees' Personal Needs.** Kelleher says he and his top managers make it a point to personally contact employees when they are ill, suffering a family crisis, or celebrating a major milestone in their lives. "We'd rather have a company bound together by love than motivated by fear."
- **Manage Well in Good Times, So That You Do Well in Bad Times.** "It is a human propensity to roll the dice in good times. But even in the best of years, we keep strong control of our costs."
- **Resist Long-Term Planning.** "It can be too confining" to a business that wants to be nimble and quick, he says. Also, resist embracing or passing on too much information. "We [in managing corporations] generate so much information of little or no relevance. A lot of time is wasted on things immaterial and hierarchical."
- **It Is More Important to Do the Right Things Than to Do Things Right.** Doing things right may still not accomplish your objectives," Kelleher says. The "right things" for Southwest include striving to offer the lowest fares in every market it serves, to be on time with every flight, to have baggage ready for travelers within 8 minutes of their arrival, and to deliver more smiles per customer than any other airline. "Amenities are frills," he says. "Our attitude is, if you want to get your baggage on time, fly Southwest. If you want to get a mint, fly someone else."

Both Jack Welch and Herb Kelleher come from very different backgrounds, have diametrically opposing personalities, and have vastly different educations. Yet the recurring themes of acquiring the best people, minimizing bureaucracy, remaining nimble, and "shooting straight" are common to both philosophies.

Not a single one of Jack Welch's 10 secrets was learned in school. Likewise, none of Herb Kelleher's rules can be traced back to law school, yet there is a common thread between the two that defines good management. What engineers need to realize is that good management does not come any more naturally to them than to a person with a business, law, or finance degree.

So, back to the question–what do engineers need beyond their natural talents and technical education to become managers? Engineers, because of the rigor of their curriculum, tend to be well above average in natural intelligence, but there are general traits that often need work.

People who become engineers are usually motivated by similar factors and have commonality in a few arenas. First and foremost, they probably did well on the math and science sections of the SAT and ACT. More often than not, they squeaked by on the language skills. Some disciplines attract the "theoretical" engineer (electrical engineering); other disciplines attract the more "hands on" types (mining). All disciplines have the same goal, to use science to solve problems. Whether the problem is to split the atom or hold a door open, the engineer will grab a pencil, some graph paper, and a calculator first. Unfortunately, their focus on technical issues is often at the expense of communication. "Engineers can't write" is a common complaint. It's not always true, but where it is, it must be corrected. No manager can be effective if he cannot communicate.

Another issue with engineers that stems from their education is the tendency to over-study a problem. In his well-known and outstanding treatise on leadership (*A Leadership Primer*), Colin Powell says:

> *... get 40 to 70 percent of the information then go with your gut. Don't take action if you have only enough information to give you less than a 40 percent chance of being right, but don't wait until you have enough facts to be 100 percent sure, because by then it is almost always too late. Today, excessive delays in the name of information-gathering breed "analysis paralysis." Procrastination in the name of reducing risk actually increases risk.*

This doesn't come naturally to an engineer. The mathematical emphasis of an engineering education requires that all or nearly all base data be collected before an answer can be calculated. No engineering professor will pass a student who does 70% of a problem then offers an answer. But in the world of management where information is imperfect and the variables are nearly infinite, it is impractical to expect to make a decision with perfect and complete data. The engineer's tendency to want to study an issue to termination must be overcome if the business expects to react rapidly.

In the *Sloan Management Review*, Edgar Schein (1996) compares the engineering culture to the executive culture. He explains, "Engineers and technocrats of all persuasions are attracted to engineering because it is abstract and impersonal. Their education reinforces the view that problems have abstract solutions and that those solutions can, in principal, be implemented in the real world with products and systems free of human foibles and errors." In explaining the executive culture, he believes that as managers rise through the hierarchy, two factors cause them to become "impersonal." They become aware that they are no longer managing operators, but managers who think like them. They also become more distant with the workforce because of the sheer size of the operation they are commanding. Thus, the executive culture has in common with the engineering culture a predilection to see people as impersonal resources that generate problems rather than solutions. Both executive and engineering cultures view people and relationships as a means to the ends of efficiency and productivity, not as ends in themselves (Schein 1996). All businesses are people businesses and must be treated as such.

If an engineer wants to make the transition to management, he would be well served to invest some time in formal training in the world of management, especially finance. Although an MBA is helpful, it is not necessary, although on-the-job training is overrated as well. Formal training, whether by short course, executive program, or university classes, will help the potential manager's comfort level when dealing with financial disciplines. Most engineers, if asked to define a term as basic but as important as cash flow would get it wrong (no, it's not the flow of cash). In 1994 General Motors made wholesale changes at the top of its management structure, replacing a group of engineers with a group that essentially came from finance (Smith 1994). The increase in profitability with the new regime supported the contention that, in a business that employs significant capital, as mining does, comfort with the financial aspects of business is important.

The last trait that is essential in a manager-to-be is the desire to become a manager. Although this is so obvious that it should not be an issue, it often is from two fronts. The first problem comes from engineers who want the added stature and financial rewards associated with management, but they have no real interest in management. Their heart lies with pure engineering ("that's why I studied engineering instead of business"), but they force themselves into management despite their true lack of interest in the field. The second problem comes when the "Peter Principle" is evoked. An exceptional engineer is pressured by upper management to take a leadership role despite the fact that the engineer is comfortable where he or she is. The engineer reluctantly takes the position and is left in a field other than the one where his or her true interests lie. In both cases it is up to the supervising manager to recognize the situation and prevent it or correct it if it is in place. It is not a rare situation. Studies indicate that about 38% of the engineers who have transitioned into management had no particular interest in making the change. They simply drifted there as a natural progression on their careers. Another 35% state that their primary motivation in moving into management was financial (Johnson and Sargeant 1998).

In a 1997 sampling of 50 firms of the 200 finalists for recognition as *Industry Week*'s 100 Best-Managed Companies, business was the degree of choice for most of the CEOs (16). However, 15 of the CEOs had engineering degrees, and finance ran a distant third with 3 (Miller 1997).

Do engineers make good managers? Sure, if they want to be and if they are willing to make the effort. They have to recognize that an engineering degree, on its own, is not necessarily enough, and any additional effort that they put into preparing for their managing career will pay off in job performance.

REFERENCES

Collins, J.C., and J. Porras. 1994. *Built to Last*. New York: HarperBusiness.

Crainer, S., and D. Dearlove. 1999. *Business the Jack Welch Way: 10 Secrets of the World's Greatest Turnaround King*. New York: AMACOM Books.

Enbysk, M. 1995. Southwest Airlines succeeds by sticking to the basics. *Washington CEO* 6(7).

Johnson D., and A. Sargeant. 1998. Motives for transition: An exploratory study of engineering managers. *Human Resource Management Journal*.

Keirsey, D., and M. Bates. 1984. *Please Understand Me: Character and Temperament Types*. Amherst, N.Y.: Prometheus Nemesis Book Co.

A master class in radical change. 1993. *Fortune*. December 13.

Melymuka, K. 2000. Down-to-earth technology helps make Herb Kelleher's Southwest Airlines a soaring success. *Computerworld*, June.

Miller, W.H. 1997. What makes a winning CEO? *Industry Week* 246(15):92.

Payne, C.M., J.E. Jeffrey, J. Burton, and F. Nuelle. December 1995. The route to the top. *Chief Executive* 109(8):24.

Schein, E.H. 1996. Three cultures of management: the key to organizational learning. *Sloan Management Review* 38(1):9.

Slater, R. 1993. *The New GE*. Homewood, Ill.: Irwin.

Smith, D.C. 1994. If you can't count beans, forget GM. *Ward's Auto World*, June.

SECTION 3

Social/Community/ Cultural Issues

. . . .

The Importance of Community Assistance in the Mining Industry

Leonard Harris* **and Rosa H. Harris†**

ABSTRACT

An important part of doing business today in the mining industry, especially overseas, involves community assistance programs for the people who live around the mining areas. The purpose of such a program is to build mutual cooperation and respect that will allow the company and its neighbors to live and work in peace. To do this, it is necessary to form associations or foundations and work in a joint venture fashion. The authors discuss in some detail the outline for setting up such programs and the important role that women, in particular, can play in them.

INTRODUCTION

In spite of increasing negative and unwarranted pressure on the mining industry (particularly in North America), and the current low selling prices for most metals, numerous exploration and development projects are forging ahead at great speed throughout the world, especially in Latin America and Africa. They range in size from megaprojects like Antamina in Peru ($2.3 billion), to San Martin in Honduras ($27 million). They cover all major metals such as gold, silver, copper, lead, zinc, platinum, palladium, diamonds, iron ore, coal, and industrial minerals.

The demand for metals to improve man's quality of life shows no indication of decreasing in the future. In addition, there is no turning back on the need to continually improve the quality of life brought to all of us by electricity, automobiles, refrigeration, air conditioning, telecommunications, high-speed travel, and recreation facilities, among others. To keep up with increasing the quality of life and extending it to the developing countries, more and more metals and minerals must be mined to manufacture these items.

It is now universally accepted that one cannot have a mining operation that creates wealth without some benefits going to the surrounding communities. This applies equally to developed and developing countries. In the words of Harry Oppenheimer, long-time chair of Anglo American and De Beers, who passed away in August 2000: "The objective of the corporation

* Mining Industry Consultant

† Founder and President for Life, ADAMINYA

is to obtain benefits for its shareholders but in a way that it makes a real and permanent contribution to the well being of the countries in which it operates."

Thus, having a community assistance program has become part of doing business, whether the company is large or small and whether it is engaged in exploration or exploitation.

A COMMUNITY ASSISTANCE PLAN

Like everything else in life, a community assistance program requires planning, planning, and more planning. There is a right way and a wrong way to set it up and carry it out. In other words, a number of "dos and don'ts" apply.

The Number 1 don't is creating dependency and paternalism. Don't give handouts. Make it a "hand-up" program instead, as expressed by Ron Cambre, chair of Newmont Mining Corporation at the American Mining Hall of Fame Awards Presentation and Banquet (December 2, 2000, Tucson, Arizona). Right from the beginning, an atmosphere of self-help should be fostered.

Don't try to do it alone. Form a foundation or an association with joint-venture partners—but under all circumstances, make sure that the company keeps control.

Don't underestimate the power of women in the community assistance program. Women from both the company and the community should be involved. They can make a powerful contribution to the success of the program, or if not taken into consideration, they can be a powerful negative force to reckon with, especially when the welfare of their children is concerned. All mothers, whether they are the wife of the chief executive officer (CEO) or the wife of a local resident, want the same thing for their children—good health, a good education, and an opportunity to make a decent living and enjoy a productive life. A female nurse or social worker, or both, should form part of the company's community assistance team.

The program should consist of short-term projects, usually some kind of health and education assistance, and long-term projects such as training for trades, higher level education, improved agriculture and animal husbandry, and small industries, among others.

The program's mission and its objectives need to be clearly defined. The broad outline of the program should be aimed at creating friendly relationships with the community and leaving a legacy of improvement in the community members' lives.

One needs to look at the program with the head *and* the heart. If one does not have cooperative relationships with one's neighbors, they can readily shut the operation down. Many of us are well aware of or have even been witnesses to some ugly events and friction between miners and their neighbors, trouble that has led to mine shutdowns and even, in extreme cases, deaths caused by conflicts between the company and the nearby communities.

At the same time, one needs to have compassion for the less fortunate and be guided by the principle of helping our fellow man to strive for a better life.

Anywhere one goes in this world, one hears the cry for better health and education. In the United States, these were the prominent items covered by the politicians during last year's presidential campaign. Even very small exploration companies can do something in these fields. The authors have witnessed this approach work wonders with small mining operations at odds with their neighbors. It doesn't have to be an elaborate or expensive program. Giving out some school supplies, enlisting the people themselves to help with the installation of basic first-aid posts, and training some of them to administer basic health programs can suffice.

Of prime importance is to never, ever make any promises one cannot keep. In the words of former Peruvian President General Manuel Odría, one needs to work on the principle of "deeds not words." A major component of the program should be trust. It takes time and effort to build and maintain this trust. The key is to follow through and do what one says one will do.

It is also important to not raise expectations too high. It is common in developing countries for the surrounding community to immediately assume that one has a mine as soon as one begins surface sampling and drilling holes. They need to be told that the work under way may or may not lead to a mine being developed. For this reason, any community assistance program at this stage should be limited.

At Minera Yanacocha in 1992, for example, before any mine was opened or any plant was built, the local authorities requested the building of a new airport, the erection of a new covered football stadium, and the paving of several hundred kilometers of road. They were told that there was absolutely no chance that this would happen before the mine went into production and if it were to happen at any time, it would require massive support from many parties including the local and federal governments. To date, considerable work has been done on roads with Minera Yanacocha's joint venture community assistance partners but neither a new airport or a football stadium has been built.

In developing countries, the local people may also have unrealistic expectations about hiring. In most cases, the local people do not have the necessary skills for even the relatively simple tasks of truck driving or equipment operation. Although of course a training program should be introduced to provide the locals with these skills, they should be told in the meantime that it will be necessary to bring skilled people from other parts of the country or even from other countries to put the mine into production. Whatever is done to assure the locals of this necessity, there will always be resentment, especially toward expatriates because they will be given certain privileges (such as living allowances) that the locals do not receive. Some companies from the developed countries may tend to underestimate the technical and supervisory capabilities of the local people.

In many cases, not only is it unnecessary to employ so many expatriates, it is expensive. There is an old and wise saying: "Never send a boy to do a man's job." Too often in community relations, dealing with the local community is left to a social worker, who, while a professional in his or her own right, may not have the necessary authorization to make decisions on community requests for assistance. The locals want to hear what the general manager has to say and to be able to be up-front about their requests, concerns, and desires. Although delegation by the general manager is good to a certain extent and in agreement with the basic principles of management, occasional visits by the general manager to the community works wonders.

Another prime requisite is dealing directly with the community leaders. This includes schoolteachers, who wield tremendous influence on the community people through their children. The mistake is often made by dealing only through central government or local government authorities who may have been mistreating the local people for decades by promising them all kinds of help, which never materializes, in return for their votes. Of course, one cannot leave these government people out of the picture, but the primary contacts should always be the community leaders, regardless of their political or religious affiliations.

Many communities throughout the world abide by cultures that may be centuries old. Of course, we must respect these cultures to a certain extent. However, in the case of primitive cultures dominated by folk healers in which disease, malnutrition, and illiteracy are prevalent, one of the goals of the community assistance program should be to teach the benefits of education, modern medicine, and technology. To leave such communities in total ignorance of these benefits is wrong and counterproductive.

A well-known Spanish saying is *Quien pide poco es un loco.* Literally translated, this means "Who asks for little is crazy." It is common and logical for community people to ask for everything under the sun from mining companies "who are stealing our riches." One should not be overwhelmed or worried about the flood of requests. These requests need to be reviewed, rejected, approved, prioritized, or postponed. Again, one needs to be very honest with the community people and not promise them anything that cannot be fulfilled. In some cases, it is simply a matter of just saying no.

A very important component of any community assistance program is not getting into anything that the company cannot extricate itself from, especially at the end of the mine life. An example experienced by the authors was a request to pay the electricity bill of a home for the elderly, which was found to be using high quantities of electricity to cook food in a large cooking pot. The pot was changed to a gas-fired pot, which was purchased and installed by the company, dramatically decreasing the electricity bill (i.e., a one-time donation was made rather than paying the power bill on an ongoing basis).

Another example was the request for the company to set up a school lunch program. Once a company starts supplying food for such a program, it will never be able to turn back. In this case, a joint-venture effort was mounted in which the food was supplied by nongovernmental organizations; the company supplied materials and know-how on building the cooking stoves; the community people built the stoves, supplied the cooking fuel (wood), and cooked the meals; and a "Ladies Committee" donated the cooking utensils, plates, and silverware. The company, then, avoided an ongoing obligation from which it cannot extract itself.

The most important of all the dos is keeping control of the program. The way to do this is discussed later in this paper.

Continuous communication with the community is of utmost significance. The authors know of many cases in which the communities have been asked to express their desires to the company, and they have replied that they wish to know what is going on and to be continually apprised of the company's plans. Regular meetings between the community people and the company people are the best way to achieve and maintain communications.

The easiest thing in life is to take the negative attitude. If one convinces oneself that nothing can be done, then nothing will be done. On the other hand, all problems have solutions that require some effort and a positive attitude to solve them. The importance of good friendly relationships with the communities cannot be understated. Ongoing efforts to make a community assistance program work are required.

The basic rules set out above can be applied anywhere in the world. The aim, once again, is to create cooperative relationships with one's neighbors, which allows the company to operate peacefully and uninterrupted to maximize profit and improve the lives of the neighbors, without creating excessive dependency and paternalism.

THE IDEAL ORGANIZATION

At the top of the organization chart should be a Board of Governors or a Board of Trustees composed of two or three high-level company officials and one or two independent members, according to the size of the operation. This board ensures that the company always has control of the program, especially with respect to the projects to be worked on and their expenditures.

Under the Board of Governors is a Board of Directors composed of:

- The company's general manager
- The company's administration manager
- The company's director of accounting
- The company's superintendent of community assistance

Also on this Board of Directors are representatives of:

- The local community
- The local municipality
- The government ministries of health and education

And others if necessary:

- The churches
- Nongovernmental organizations
- The armed forces
- The local and national cultural institutes
- Citizen groups

- Labor unions
- A ladies association

The duties of the Board of Directors are to develop projects for possible incorporation in the program. The board should also prepare a plan and a budget to be sent to the Board of Governors for their approval toward the end of each year.

In this way, all concerned have a say in what the program should look like and all involved have an opportunity to put forward their wish lists.

THE MINERA YANACOCHA MODEL

Much has been written and published about Minera Yanacocha's gold mining operations in northern Peru and its very successful community assistance program. The program was described in some detail in the August 1999 issue of *E&MJ* in an article entitled "The Yanacocha Project." Some 30 communities are involved, which include about 50,000 people.

The model set by Minera Yanacocha has been used to replicate similar programs in Indonesia and Honduras and are actively being examined for Mexico and Turkey. In addition, visits have been made by many companies from the United States, Canada, Chile, Bolivia, and Ecuador to investigate the possibilities of employing the model elsewhere.

In accordance with the principle of not trying to do it alone, Minera Yanacocha has set up an association. The members of this association are:

- Minera Yanacocha and its partners (Newmont Peru Ltd. and Compañía de Minas Buenaventura)
- The communities
- The Cajamarca Municipality
- The Ministry of Health
- The Ministry of Education
- The Ministry of Agriculture
- The Ministry of Transportation
- The armed forces
- The churches
- The University of Cajamarca
- The Cajamarca Chamber of Commerce
- C.A.R.E. Peru
- Inter American Foundation
- The Royal Dutch Technical Cooperation Agency
- The World Food Program–United Nations
- INCALAC (Nestlé)
- PRONAA (Peruvian Program for Providing Food for the Poor)
- SENATI (Peruvian National Trade School)
- ADEFOR (Forestry R&D)
- FONDER (Regional Development Program)
- Fondo Empleo (Employee Training Fund)
- National Institute of Culture
- Ladies Association Minera Yanacocha
- Providers of goods and services

The program carried out by Minera Yanacocha in conjunction with its joint venture partners consists of various projects, as follows:

Health Projects–First-aid posts, training of paramedics, nutritional education, immunization, prenatal care, training of midwives, installation of potable water systems, and construction of latrines.

Education Projects–New schools, rehabilitation of existing schools, school supplies, school lunch programs, and school garden plots.

Agricultural Assistance Projects–Improvement of pastures, tree planting, grain and seed storage facilities, animal husbandry, and irrigation ditches.

Road Improvement Projects–New rural roads, in addition to upgrading and maintaining existing roads.

Electrification Projects–Small power plant studies and extension of Minera Yanacocha's main power line into the villages and hamlets.

Small Business Projects–Credit assistance for small sawmills, sewing and weaving enterprises, stone carving shops, and trout farming.

THE LADIES ASSOCIATION

An extremely important partner in the Minera Yanacocha Community Assistance Program is the Ladies Association Minera Yanacocha, known by its Spanish acronym ADAMINYA. This association was formed in October 1993 by 12 women, including wives of Minera Yanacocha's senior staff members and single female employees of the company. Today, the association has more than 200 members.

Rosa Harris was elected as the association's first president and today remains its founder and president for life. She refers to the overall program managed by Minera Yanacocha as the muscle of the program, with the Ladies Association as the heart of the program. In other words, men mostly think with their heads whereas women tend to think mostly with their hearts in reference to community assistance programs.

The mission of the Ladies Association is "to give the children who attend the schools around Minera Yanacocha's mining operations the opportunity for progress, so they may improve their lives and have control of their future." The programs carried out by ADAMINYA are as follows:

Education–A school supplies campaign was started in 1994 in 14 schools with 900 students. In 2000, 33 schools are covered with 4,000 students. This campaign consists of donating textbooks, notebooks, pens, pencils, crayons, desks, tables, chairs, and typewriters.

School Lunch–This joint venture program was also started in 1994 as a pilot program together with Minera Yanacocha and C.A.R.E. covering 300 children in 5 schools. In 2000, 23 schools are covered with 3,000 children. ADAMINYA's part of this program is the supply of cooking and eating utensils.

Mother/Child Health–In the rural areas of Peru, childbirth usually takes place in homes where professional assistance is difficult to reach. ADAMINYA has organized courses in midwifery for the community women. Such courses were initiated in 1995, and by 2000, 23 women and one man have graduated. The course lectures are given by medical professionals from the School of Obstetricians in Cajamarca City.

Health Program– In its desire to protect the health of the children, ADAMINYA established a permanent campaign aimed at promoting personal healthy habits by the school children. The ladies of ADAMINYA work together with the schoolteachers and the parents on dental care, lice control, and treatment of scabies.

School Garden Plots and Vocational Training–The school garden plots project is aimed at augmenting the food for the school lunch program and instilling the concept of self help, thereby becoming less dependant on donated food. The results have been nothing less than spectacular. The children have taken the idea to their homes where they and their parents have planted garden plots. Vocational study covering sewing, food preparation and processing, tool making, carpentry, and typing is the next area under study.

Fund Raising–The ladies of ADAMINYA, under the direction of their Board of Directors, carry out a series of fund-raising events throughout the year to obtain the funds needed for their projects. These consist of the sale of Christmas cards, raffles, barbecues, organization of Minera Yanacocha's annual Family Day Picnic, preparation of Minera Yanacocha's Christmas baskets, and a dinner/dance fundraiser every year in Denver.

ADAMINYA also receives significant donations from various companies, which supply goods and services to Minera Yanacocha and individuals, and receives a matching gift donation every year from Minera Yanacocha.

A list of providers of goods and services who donated to ADAMINYA's cause in 1999 follows:

- Baker Hughes Mining Tools Peru S.A.
- Bechtel Corporation
- Boart Long Year
- Ferreyros S.A.
- Forza S.A. Seguridad Privada
- Geotec S.A.
- Geotorno
- Hermes Transportes Blindados
- Knight Piesold & Company
- Lemerox S.R.
- Odebrech
- Perforaciones St. Lambert
- Pumps Plus Inc.
- S.G.C. del Perú S.A.
- V.S.V. Contratistas Mineros
- Water Management Consultants (Peru) S.A.

To summarize, the success of a community assistance program depends on:

- Trust
- Self help
- Cooperation

The results obtained to date clearly demonstrate the attainment of these important items by ADAMINYA at Minera Yanacocha.

CONCLUSION

The success of a community assistance program at mine sites depends on the application of basic principles–planning, organization, delegation, and control. In addition, a structure should be set up under which the company keeps control of the program and its expenditures. Involving the people of the community along with the company's partners, its contractors, the local governmental and nongovernmental organizations, and the providers of goods and services, is key. The programs should be aimed at promoting friendly relationships with the mine's neighbors without creating excessive dependency or paternalism. These positive relationships will allow the company to operate peacefully and continuously, maximizing profit and improving the lives of those in the surrounding community.

. . . .

International Lender Finance of Mining Projects

Environmental Requirements for Prefeasibility and Feasibility Studies, and for Project Implementation

Rick L. Williamson*

ABSTRACT

International lending institutions have charters requiring projects that are seeking loans to adhere to certain environmental performance standards, including compliance with host-country environmental laws, as well as observance of international environmental policies and guidelines. The most commonly applied international environmental policies and guidelines are those established by The World Bank. Lenders require a review by an independent engineer to determine that the feasibility study plans have been developed in accordance with the applicable environmental commitments. Lender agreements require independent engineer evaluation of environmental performance through the construction phase and initial operations phase of the project, culminating in the issuance of an "Environmental Completion Certificate."

INTRODUCTION

Commercial banks and government-export credit agencies (international lending institutions) have charters requiring projects that are seeking loans to adhere to certain environmental performance standards as a key part of the financing agreement for the project. These requirements include compliance with the host-country environmental laws and regulations, as well as observance of international environmental policies and guidelines. In Pincock, Allen & Holt's (PAH) experience, The World Bank environmental policies and guidelines have generally been the international lending institutions' criteria of choice. However, the reader should be aware that other environmental guidelines are also used at times, either in concert with The World Bank guidelines, or as the sole criteria for the project. Other environmental guidelines include those provided by The European Investment Bank (EIB), the Export-Import Bank of the United States (USEXIM), and those established by the International Finance Corporation (IFC). The primary aim of EIB's guidelines is to verify that prospective finance projects are in compliance with prevailing European Community or national environmental legislation and standards, or both. The USEXIM guidelines are generally used in cases where USEXIM is a

* Environmental Specialist, Pincock, Allen & Holt

senior lender. USEXIM's environmental guidelines are similar to those established by The World Bank. The IFC guidelines are designed to supplement those established by The World Bank. Other guidance documents, such as those promulgated by the European Bank for Reconstruction and Development (EBRD), as well as the Canadian Export Development Corporation's (EDC) Environmental Framework Document, are also in use.

Documents addressing The World Bank's environmental policies and guidelines can be found in a number of published reports; however, most of these requirements are addressed in:

- *The World Bank Pollution Prevention and Abatement Handbook 1998* (April 1999). The environmental policies and guidelines discussed in this document apply to coal, base metal, and iron ore mining operations.
- *The World Bank Policies and Guidelines: Environment, Health and Safety Guidelines: Mining and Milling* (August 1995). These environmental guidelines are the most current for application to precious metals mining operations.
- *The World Bank Environmental Assessment Sourcebook* (most recent version published in October 1996, with subsequent updates published to clarify subject-specific environmental policies).
- *The World Bank Operational Directives/Policies.* These operational directives/policies (ODs/OPs) address specific subjects, and include ODs 4.00 and 4.01 (Environmental Assessments), OD 4.20 (Indigenous Peoples), OD 4.30 (Involuntary Resettlement), OP 4.04 (Natural Habitats), OP 4.07 (Water Resources Management), OP 4.37 (Safety of Dams), and OP 11.03 (Management of Cultural Property).

Lending institutions require that project plans undergo a detailed due diligence review by an independent engineer (IE), to determine whether these plans have been developed in accordance with the applicable environmental policies and guidelines for the project seeking the lender's financial involvement. This review is a common requirement for feasibility studies, and extends to prefeasibility studies as well, if, as has been the case in a few small-scale projects, the lenders use this latter document as the primary instrument upon which the loan agreements are based.

Lender agreements usually require IE evaluation of environmental performance through the construction phase and the initial operations phase of the project, culminating in the IE's issuance of a certification document indicating that the project plans have been designed and completed in compliance with the applicable environmental commitments initially provided to the lenders.

PREFEASIBILITY STUDIES AND PRELIMINARY ENVIRONMENTAL BASELINE STUDIES

From an environmental perspective, the primary focus of a prefeasibility study document is to accomplish the following:

- Identify environmental physical factors that may have a significant bearing on the construction and operational phase design plans for the project. These factors may ultimately affect project costs and feasibility. These factors may be associated with, for example, the climatic conditions of the site, the presence or absence of surface water or groundwater, the nature of the site terrain, the existence of a suitable local labor force, and predicted seismic conditions.
- Identify potential environmental fatal flaws that may be present that could cause serious difficulties in permitting the operation through the local governmental regulatory authorities, or may cause environmental impacts that cannot be mitigated to an acceptable degree, because of excessive costs or other physical factors.

- Identify environmental liabilities that may already exist on the property, which could significantly elevate the environmental costs associated with the new operation.

The above determinations can be realized through the preparation of a preliminary environmental baseline study for the site. This study should be the precursor to the more detailed work to be conducted for the feasibility study and should, at a minimum, address the following subject areas:

- Surface water hydrology
- Climate and meteorology
- Groundwater hydrology
- Socioeconomics
- Terrestrial and aquatic flora and fauna
- Land and water use
- Archaeological and cultural resources
- Presence of indigenous peoples
- Air quality
- Need for relocation of population
- Marine resources (if present)
- Conflicts with international agreements
- Noise and vibrations
- Public sentiment
- Topsoil resources
- Potential for acid rock drainage

It should be emphasized that determinations included in a prefeasibility study are generally considered to be preliminary in nature.

FEASIBILITY STUDIES AND ENVIRONMENTAL IMPACT ASSESSMENTS

Feasibility studies differ from prefeasibility studies primarily in the level of detailed analysis required. Feasibility studies must include the preparation of a detailed environmental baseline study and a subsequent environmental impact assessment (EIA) for the project. The EIA should be prepared in accordance with the environmental impact assessment policies established by The World Bank, or other established acceptable standards. In the case of The World Bank, these policies are addressed in ODs 4.00 and 4.01, and are further discussed in the bank's environmental assessment sourcebooks.

A checklist of potential issues for consideration during preparation of an EIA would include:*

- *Biological Diversity*: Particularly in regard to endangered plant and animal species, critical wildlife habitats, and protected areas
- *Coastal and Marine Resources Management*: Especially coral reefs, mangroves, and estuaries
- *Cultural Properties*: Protection of archaeological sites, historic monuments, and historic settlements
- *Dams and Reservoirs*: Environmental issues in planning, implementation, and operation of projects that include large dams and reservoirs

* An adaptation of the list included in World Bank Operational Directive 4.01.

- *Hazardous and Toxic Materials*: Safe use, transport, storage, and disposal
- *Indigenous Peoples*: Traditional land and water rights
- *Induced Development and Other Sociocultural Aspects*: Secondary growth of settlements and infrastructure (i.e., "boomtown" effects)
- *International Treaties and Agreements on Environment and Natural Resources*: Status of project effects on current and pending treaties and agreements
- *Involuntary Resettlement*: An especially sensitive issue to many lenders, particularly if and when indigenous peoples are involved
- *Natural Hazards*: Potential effects to the project that may be caused by earthquakes, floods, volcanic activity, among others
- *Occupational Health and Safety*: Formal plans to promote safety
- *Project Support Facilities such as Ports, Power Supply, and Access Roads*: Environmental considerations for management through the construction, operations, and closure phases
- *Industrial Hazards*: Management of industrial hazards
- *Watersheds*: Protection and management of watersheds affected by the project
- *Ports and Harbors*: Associated environmental concerns where applicable
- *Wetlands and Wildlands*: Particularly estuaries, natural lakes, mangroves, marshes, swamps, pristine forests, and other wildlands under special protection mandates
- *Land Settlement*: Associated impacts
- *Public Consultation*: A key project component

An outline for an acceptable EIA report would include the following primary sections:*

- *Executive Summary*: Presents findings and recommended actions
- *Policy, Legal, and Administrative Framework*: Describes the framework within which the EIA was prepared
- *Project Description*: Outlines the geographic, ecological, social, and temporal context of the project
- *Baseline Data*: Presents the dimensions of the study area and description of relevant physical, biological, and socioeconomic conditions present
- *Environmental Impacts*: Covers the anticipated impacts as a result of project implementation
- *Analysis of Alternatives*: Systematically compares the proposed investment design, site, technology, and operational alternatives in terms of their potential environmental impacts; capital and recurrent costs; suitability under local conditions; and institutional, training, and monitoring requirements
- *Impact Mitigation Plan*: Identifies feasible and cost-effective measures that would likely reduce potentially significant adverse environmental impacts to acceptable levels
- *Environmental Management and Training Assessment*: Identifies the existence, role, and capability of environmental units on site to determine the necessity of recommendations to establish or expand such units, or both, and the training of staff, to the point that EIA recommendations can be implemented
- *Environmental Monitoring Plan:* Specifies the type of monitoring, the entity that would conduct the monitoring, and costs

* An adaptation of the list included in World Bank Operational Directive 4.00.

PROJECT CONSTRUCTION, OPERATIONS, AND MINE CLOSURE REQUIREMENTS

World Bank environmental guidelines specifically discuss the need for project development plans to address:*

- The removal, proper storage, and management of topsoils
- Early restorations of worked-out areas and of spoil heaps to minimize the extent of open areas
- Identification of potential areas for acid rock drainage (ARD) generation
- A water management plan focusing on the effective use of mine water for operations and for postclosure
- Extraction methods in relation to subsidence and to surface use
- Development of restoration and vegetation methods that are appropriate to the specific site conditions
- Blasting methods that minimize noise and vibrations
- The minimization of erosion and sedimentation
- Tailings disposal plans that optimize human safety and environmental protection
- Mine closure and reclamation plans that include:
  - Return of the land to conditions capable of supporting prior land use, equivalent uses, or other acceptable uses
  - Elimination of significant adverse effects on adjacent water resources
  - Use of waste rock for backfill and of topsoil (or other acceptable materials) for reclamation to the extent feasible
  - Contouring of the waste dump slopes to minimize erosion and runoff
  - Planting of native species of vegetation and of other species that are environmentally acceptable, to prevent erosion and to encourage self-sustaining development of a productive ecosystem on the reclaimed land
  - Postclosure management of ARD and tailings; reduction of ARD formation by sealing off pyrite-containing waste from oxidation and percolating water
  - Sealing or securing of all shaft openings and mine adits on closure of the mine
  - A budget and schedule addressing costs for pre- and postclosure reclamation activities

The World Bank environmental policies and guidelines for mining projects also include specific emission standard recommendations for liquid effluents, air quality, and noise. In addition, these guidelines address monitoring and reporting frequencies for project emissions.

* Please note that the issues listed are not a comprehensive/complete listing of the recommended measures addressed in The World Bank environmental guidelines.

. . . .

Nongovernmental Organizations: Friend or Foe?

James P. Cooney*

ABSTRACT

The global mining industry is in the process of redefining its activity and purpose in terms of sustainable development. This process is impelling a growing number of companies to engage their stakeholders on environmental and social issues. Nongovernmental organizations (NGOs) have customarily acted as critics of the mining industry's performance. A new relationship between mining companies and NGOs is emerging in the context of sustainable development, ranging from dialogue to joint action. If positive interaction among mining companies and NGOs is to succeed, clear terms of engagement must be articulated, and measurable progress toward achieving both corporate and NGO objectives must be demonstrated. The experience of Placer Dome is illustrative of the rewards and challenges of engaging with NGOs.

INTRODUCTION

Globalization has opened a Pandora's box of conflicting values, objectives, and priorities. Multinational corporations and a number of governments are focused on the wealth creation made possible from accelerated international trade and investment. Many churches and nongovernmental organizations (NGOs), on the other hand, are primarily concerned about alleviating poverty through the redistribution of wealth and the forgiveness of developing country debt. They see trade and investment as worsening the income gap between the rich and the poor.

For some groups and organizations the primary issue of globalization is that the decisions made by international institutions, governments, and corporations lack transparency. They worry about hidden agendas and inappropriate influence on decisions that affect the common well-being. For some NGOs the principal issue is that individual and collective human rights are not recognized and enforced in many jurisdictions of the world. They argue that real economic and social progress in the world is not possible unless and until fundamental human rights are respected globally. Some maintain that education and health are the fundamental prerequisites for human progress. They would make the eradication of disease and the expansion of literacy and numeracy the global priority.

For many groups and organizations the protection of the environment is the indisputable global priority. They are particularly concerned about the issues of biodiversity and climate

* General Manager, Strategic Issues, Placer Dome, Inc.

change. Other organizations, particularly in the trade union movement, see the establishment of core labor standards as the global priority. They are concerned that the generation of wealth through international trade and investment is sometimes at the expense of child or forced labor, and often without an adequate recognition of the freedom of association and collective bargaining. Many communities, particularly those designated as indigenous, are concerned about the preservation of traditional culture. They see the intrusion of multinational corporations into their vicinity as a threat to their languages, customs, and mores. Finally, many religious people view the entire process of globalization as a materialistic thrust, which distracts from and even impedes the spiritual life of human beings.

HOW NONGOVERNMENTAL ORGANIZATIONS VIEW CORPORATIONS

How do multinational corporations relate to this plethora of global issues? Many international NGOs, academics, and church leaders would say not well. They question whether multinational corporations are responsible. They accuse corporations of fundamentally not caring about the environment, human rights, labor standards, community well-being, or national sovereignty as much as they care about business success and the relentless pursuit of profit. What they seek from corporations are ironclad commitments to protect the environment; promote human rights; respect labor standards; benefit communities; and comply with national laws, regulations, and court decisions.

NGOs question whether multinational corporations are accountable. They accuse corporations of answering only to their shareholders, and not to all the other stakeholders that are affected by corporate decisions and actions. They demand that corporations respond and report to stakeholders on matters of concern in a credible manner. Many NGOs demand that corporate sustainability performance be subject to independent third-party monitoring and reporting.

NGOs also question whether multinational corporations are capable of ethical decision making. Frequently, they quote the celebrated dictum of Milton Friedman, that the ethical obligation of a chief executive officer (CEO) is to provide the greatest possible return to shareholders, subject only to the limits of legal statute, public perception, and moral custom. They ask, then, how a corporation can possibly integrate into its decision processes considerations of social, distributive, and procedural justice; of the common good and human rights; and of the values of a virtuous society. They seek either to radically restructure corporate decision making so as to include these ethical factors, or to deprive corporations of decision-making autonomy, subjecting them to regulation by a higher authority dedicated to the common interests of society.

THREE PERSPECTIVES ON GLOBALIZATION

The question facing society globally is how to structure, integrate, and harmonize all these competing demands for attention. Currently, three policy frameworks compete as organizing structures for the full range of global issues and human concerns: sustainable development, human rights, and human security.

Sustainable Development

Conceptualized by the Brundtland Commission and launched at the United Nations Summit at Rio de Janeiro in 1992, sustainable development has been adopted by UN agencies, the World Bank, and other multilateral institutions, as well as by some national governments and NGOs, as the preferred global policy framework. Sustainable development structures global issues in environmental, social, and economic categories, which need to be integrated and balanced in governmental and corporate decision making, in a way that uses the resources of the present with a respect for the needs of future generations.

Human Rights

As articulated in the United Nations' Universal Declaration, and the two conventions on political and civil rights and on economic, social, and cultural rights, the concept of human rights has been adopted by some international agencies, NGOs, and churches as the preferred global policy framework. The human rights paradigm structures global issues in terms of certain absolute entitlements that every human being should enjoy continuously (such as freedom from torture or from forced labor), along with certain progressive entitlements that governments must work to implement over time (such as full employment or universal health care).

Human Security

In 1994, the United Nations' Development Program defined human security as embracing economic, food, health, environmental, personal, community, and political dimensions. The concept structures global issues in terms of threats to human well-being. Decision processes in the paradigm of human security are focused on protecting the lives and livelihoods of individuals, the integrity of communities, and the sovereignty of nations, recognizing that the security of the state is not an end in itself, but a means of ensuring security for its people.

THE CORPORATE EMBRACE OF SUSTAINABLE DEVELOPMENT

In February of 1998, the Placer Dome Board of Directors adopted a corporate sustainability policy. Of the three policy frameworks, sustainable development conforms most closely to the corporate way of understanding issues and making decisions. Sustainable development envisages collective action to achieve progressive outcomes for human betterment. The engineering mind-set of a company like Placer Dome similarly employs consequentialist logic to design and implement improvements of an operational sort. Industrial corporations view sustainable development as a way to engineer improvements in society.

The human rights and human security paradigms are less amenable to corporate thought processes. Rights appear to be easy to assert, but difficult to demonstrate, and hard to measure in any engineering or quantitative frame of reference. Security seems to relate to the context of human and societal self-actualization rather than to human progress itself. Establishing a context of security within which human development can occur is, for corporations, more a governmental than a corporate agenda.

It is important, therefore, to recognize that while Placer Dome, along with other corporations, multilateral agencies, and NGOs, has adopted the language and logic of sustainable development, not everyone understands or thinks about issues in the same terms. At times, for example, the author has entered an aboriginal community in Canada to discuss a proposed project, with the overture: "Let's talk about jobs and benefits and environmental protection." But the response has been: "Let's talk about our rights, our right to decide, our right to resources under our land, our right to determine our own future." Engagement with stakeholders is difficult enough when we and they share a common set of fundamental principles, such as sustainable development, whose implications and implementation are all that we need to agree upon. Engagement is doubly difficult when the interested parties understand the world and think about issues using different concepts and different logical connections.

REASONS FOR ENGAGING STAKEHOLDERS

For any corporation that embraces sustainable development, engaging stakeholders is essential. Placer Dome formulated its sustainability policy through considerable dialogue with external stakeholders (international NGOs and multilateral institutions, such as the World Bank) and with internal stakeholders (employees). Action plans to achieve the commitments of the sustainability policy are drawn up by each Placer Dome operation in recognition of the priorities of its local stakeholders. The implementation of sustainability action plans normally involves stakeholders, either in a consultative role or sometimes as partners in delivering social programs. Finally, the assessment of progress and shortfalls in achieving sustainability commitments requires the advice of stakeholders on benchmarks of performance and credibility in reporting. Placer Dome has issued two global sustainability reports since adopting its sustainability policy, and each has involved considerable consultation with international NGOs and other stakeholders. Currently, Placer Dome is exploring a possible partnership with the World Wildlife Fund (WWF) to develop a program to certify the sustainability performance of a mine.

A corporation should engage its stakeholders for several reasons. First, legislation in certain jurisdictions, such as Canada, may prescribe stakeholder engagement as part of the environmental assessment and permitting processes. Second, some stakeholders have considerable influence on government decisions about project proposals. Third, corporations stand to benefit from more cost-effective and less risk-prone decisions, insofar as stakeholder consultation produces more complete information about expectations, sensitivities, and capacities for change. Finally, it would appear that individuals and communities have a right to be involved in decisions that affect them.

The potential contribution of stakeholders to corporate decisions and actions is considerable. Stakeholders often supply a certain credibility that corporations lack. Any survey of public opinion will show that the greatest credibility attaches to NGOs, academics, and religious and community leaders. Medium credibility is enjoyed by public servants and officials of major institutions, such as United Nations (UN) agencies. On the other hand, corporate leaders are ranked low, along with politicians, in terms of credibility. The reason is obvious. Business people and politicians are perceived by the public to be extremely self-interested, whereas those with high credibility appear to be concerned, if not about the common good, about some cause larger than their individual self-interests. Consequently, corporate alliances with NGOs, academics, and religious and community leaders around issues of common concern will give a corporation's decisions and actions greater credibility with the public.

Stakeholders can also contribute knowledge to the corporate understanding of issues. Stakeholder awareness of local circumstances, relationships, values, and priorities can significantly broaden a corporation's knowledge base for critical decisions. Stakeholders also possess competencies related to implementing sustainability strategies, which corporations often do not possess. For example, NGOs have expertise in aspects of environmental protection such as biodiversity conservation, in rural health and education, in microcredit financing, and in community capacity building. Alliances with such NGOs can broaden the capability of corporations in the delivery of sustainability programs in the field.

Certain stakeholders also have access to financial and other resources (such as their communications networks) that can expand the support system for corporate sustainability programs. Donor nations provide considerable financial support to NGOs. Such public funding is not normally available to corporations, however meritorious their sustainability programs. Corporate sustainability initiatives around operations may lever foreign aid through NGOs into applications that complement corporate programs. Ultimately, sustainable development, given the dimensions of the challenge in many parts of the world, cannot be achieved by a mining company acting on its own. It requires partnerships and burden-sharing with NGOs and bilateral donors.

DEFINING STAKEHOLDERS AND THEIR INTERESTS

Stakeholders can be broadly defined as any person or group that is affected by or concerned about a corporate activity (see Figure 1). NGOs, of course, are only one set of stakeholders, and generally not the most important ones. Local communities are the people who are always the most affected by and concerned about a mining project. Other local groups, such as churches, schools, hospitals, and special interest organizations, are also important

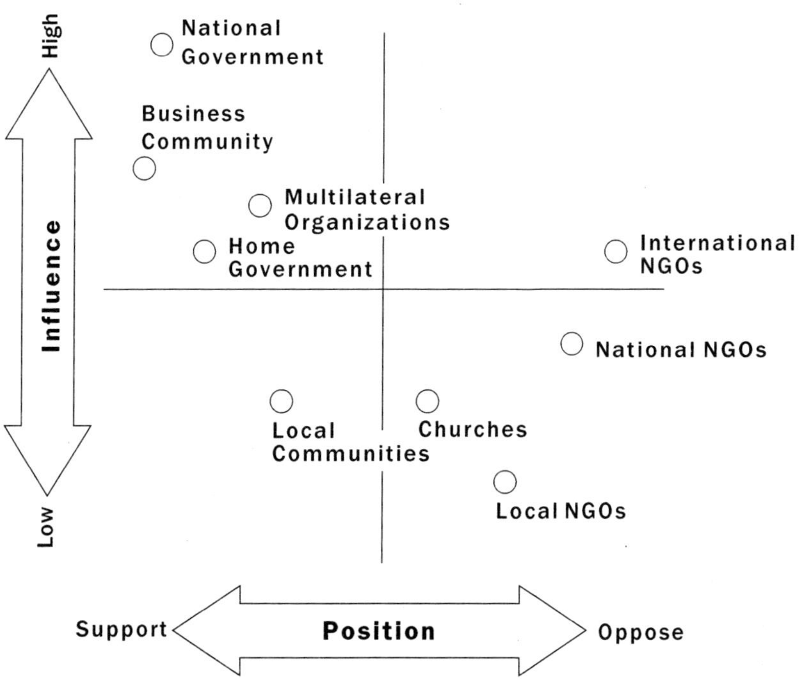

FIGURE 1 Typical positions and influence of stakeholders: mine development

stakeholders. Governments at all levels can also be considered stakeholders to a corporation, although governments may see themselves as decision makers with their own stakeholders in the form of communities, NGOs, and other institutions. Finally, multilateral institutions, such as UN agencies and the World Bank, international NGOs, and the home-country government of an investing corporation are also stakeholders.

Generally speaking, stakeholder issues can be clustered in three categories. Stakeholders are concerned about the environmental and social impacts of a project. They seek assurances that negative impacts will be prevented if at all possible; if not preventable, that impacts will be mitigated insofar as possible; and where mitigation is inadequate, that compensation will be paid for such impacts.

Stakeholders are also concerned about the benefits from a mining project that will flow to individuals and communities in the local area and to the country overall. They want to maximize the availability of jobs, business opportunities, social and physical infrastructure, and direct revenues from the project to local coffers and the national treasury.

Finally, stakeholders are concerned about individual and collective rights (see Figure 2). Local communities may assert their exclusive power to decide if a project proceeds, their extensive entitlement to benefits from a project in their vicinity, and their absolute freedom from manipulation by governments or foreign corporations.

FINDING COMMON GROUND

It should be evident that it is not easy for corporations to find common ground with all stakeholders, particularly with NGOs. Moreover, not all NGOs are of the same mind with respect to how they can or should relate to corporations. At one end of the spectrum are NGOs who are so radically opposed to corporations and the free market economy in general that dialogue, much less collaboration, between them and corporations is virtually impossible.

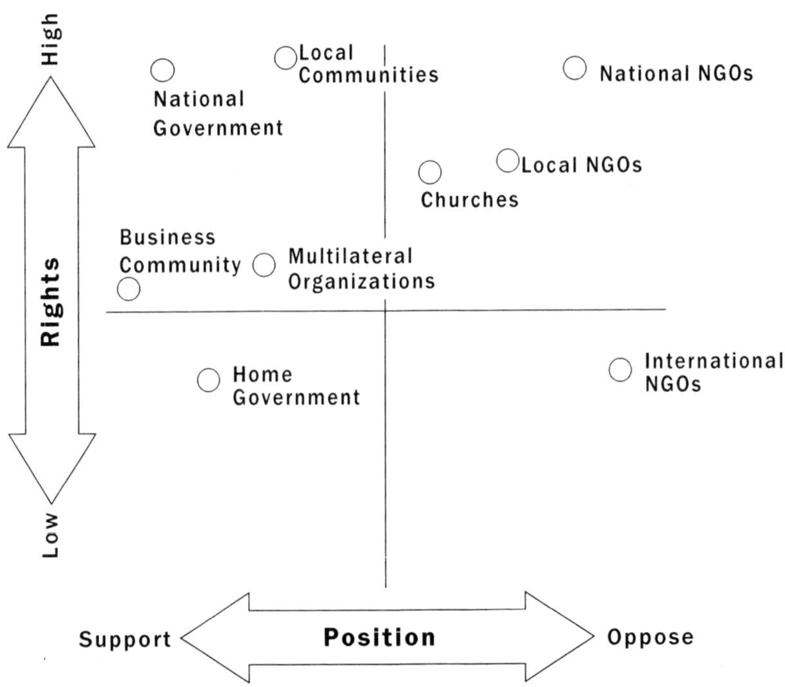

FIGURE 2 Typical positions and rights of stakeholders: mine development

At the other end of the spectrum, however, are NGOs who take a constructive approach to working with corporations. Such NGOs seek opportunities to engage corporations around issues of common concern, to develop shared solutions, and even to work together in the field to implement sustainability programs. CARE International, which cochairs the Natural Resource Cluster of World Bank's Business Partners in Development Program, is an example of a constructivist NGO.

In the middle of the spectrum are activist NGOs, such as WWF, Amnesty International, and the Mineral Policy Centre, which will opt for confrontation or engagement with corporations, depending on which tactic seems most likely to be effective under the circumstances. Corporations that seek to engage a broad base of NGOs as part of their stakeholder engagement strategy need to find ways to attract activist NGOs into dialogue and possibly collaboration, and away from confrontation and conflict.

Stakeholders differ in their relative influence in corporate and government decisions about a project. Some stakeholders, with a relatively high level of influence on project decision-making, are generally favorable to project proposals. The national government and the domestic business community of a host country have a lot to gain financially from major investments in mining in their country, and so they are normally quite favorably disposed toward project proposals. Multinational organizations, such as the World Bank, regional development banks, and the UN agencies are generally supportive of new investment in mine development, as long as certain safeguards with respect to sustainable development (environmental and social impacts) are in place. These organizations may also exert considerable influence on the decisions of host governments. Finally, the home government of the investing corporation, such as Canada in the case of Placer Dome, is generally supportive of new mining proposals for the trade linkages they encourage. Such home governments may have some influence on decisions by the corporation and by the host government.

On the other hand, local communities, while generally supportive of new mining projects in their vicinity for the benefits they offer, often have little influence on host governments or corporations. There are, of course, exceptions to this norm, as in the case of aboriginal peoples in Canada and Australia and local landowners whose rights are constitutionally protected (as in Papua, New Guinea).

There are other stakeholders who are often hesitant, skeptical, and opposed to new mine developments because of their concerns about social and environmental impacts. They include local churches and local or national NGOs. However, their influence on government and corporate decisions is generally slight. On the other hand, some international NGOs, which generally oppose mining projects for a variety of reasons, may through their media and institutional networks be able to exert a certain amount of influence on corporate and government decisions.

Corporations tend naturally to relate most comfortably to those stakeholders who have significant influence on project decisions and are also highly influential. The tendency, therefore, is to pay scant attention to local churches and local NGOs. However, despite the relative lack of influence they generally have on governmental decisions, local communities often attract a considerable degree of corporate attention, because the development and operation of a mine often require some measure of local cooperation.

International NGOs are the wild card in mine development decisions. To corporations, their opposition to projects often seems unreasonable and extreme. Yet the larger international NGOs exert some degree of influence on multilateral institutions, such as UN agencies and the World Bank, as well as on some home and host governments. Increasingly, mining companies are finding that engaging large international NGOs who are willing to enter into dialogue with an objective of finding some measure of common ground—perhaps of integrating some NGO concerns into corporate decision-making, and certainly of softening NGO opposition to a given project—is a key element to a corporate strategy for winning government approval and obtaining the necessary permits.

RECOGNIZING STAKEHOLDER RIGHTS

The motivation for consulting stakeholders is different if viewed from the perspective of rights, instead of influence. It can be argued that local communities, although their influence on decisions may be limited, have the most significant right to be involved in decision making. Similarly, local churches and local and national NGOs, given their moral authority, their interest in the common welfare, and the extent to which they represent the values and priorities of host country nationals, may also claim a right to be consulted.

The author recalls his various conversations with local church leaders about proposed mine developments in rural areas of Latin America and Africa. As the project proponent, he would describe the jobs, business opportunities, and other revenue flows that the communities in the vicinity of the project might expect. His clerical interlocutor—a minister, a priest, or a bishop—might acknowledge the potential benefits, but then add: "These rural people live largely in a cashless society. When you start paying wages and pumping money into their villages, and the result is gambling, prostitution, drunkenness, family violence, and the general disintegration of the community, who will take responsibility? What will you do?"

The corporate answer has been: "We need partners. We need to work with organizations like churches to shore up community values and morality. Certainly, we cannot take on total responsibility for the actions of individuals or the well-being of the community." The rejoinder sometimes is: "What right do you, as a corporation have to inflict such adjustments, the evils of the wage economy, on these poor people?"

One needs to consider whether the right of the community to maintain its integrity, as well as the right of churches and other groups to defend the well-being of the community, may be superior to the rights of an investing corporation. Understandably, corporations will seek to shift the conversation away from rights toward sustainable development, where the objective is to engage stakeholders in shared decisions and possibly joint action with a view to achieving collective economic, social, and environmental progress.

It is worth observing that the rights as stakeholders of the home government and of international NGOs are relatively insignificant. Corporations relate to them more because of their influence on decisions than because of their rights to be involved.

FORMS OF STAKEHOLDER ENGAGEMENT

At one level of strategy and action plans, engagement with stakeholders can take a number of forms. Education is essentially a matter of the corporation communicating its plan to potentially affected communities and organizations. Feedback is the receiving by the corporation of input from stakeholders about their interests and concerns. Education and feedback are relatively weak forms of engagement between a corporation and its stakeholders.

A more significant form of engagement in the development of corporate strategy and action plans is consultation, or two-way dialogue. Sometimes the dialogue evolves into joint planning by the corporation and stakeholders, where major decisions on objectives and actions are reached by consensus. Sometimes, the differences between corporations and stakeholders are too difficult to resolve without professional facilitation of this dialogue and decision process. Then objectives and actions may be agreed on through mediation or dispute resolution. Moving from consultation, through joint planning, to mediation, the corporation experiences diminishing degrees of control over the dialogue and decision process, and increasingly shares control with stakeholders.

When it proves impossible to resolve disputes between a corporation and stakeholders through mediation, it may be necessary to resort to arbitration. When arbitration fails, the last resort is litigation. In both arbitration and litigation, the corporation loses all control over the decision process. Generally speaking, this is not how corporations want decisions about projects to be made.

At the level of implementation and reporting related to sustainability objectives, there are also different levels in the intensity of engagement between a corporation and stakeholders. The parties may simply engage in consultation on their various programs and activities, observing complementary or conflicting areas. They may decide to align their programs and activities to make them complementary. They may go further and collaborate informally in certain programs and activities, or they may even make the collaboration formal through a contractual relationship. The ultimate form of acting together in implementing programs or reporting on progress is to assign responsibility to a single agency that acts on behalf of the corporation and stakeholders, a form of joint delivery of the contribution to sustainability.

STAKEHOLDER ENGAGEMENT FROM EXPLORATION TO MINE CLOSURE

Corporate objectives for stakeholder engagement will vary according to the stage of a project. The exploration stage is the time of first contact between a corporation and stakeholders, be they local communities, NGOs, or governments. The corporate objectives at the exploration stage are to gain and secure access, to establish relationships, to identify issues, and to manage expectations.

At the feasibility study stage, when mine planning and permitting occur, the corporate objectives are (1) to identify and document environmental and social issues that may affect project design and permitting; (2) to develop public support for the project and diffuse public opposition by incorporating the public consensus into project design; and (3) to satisfy government-mandated consultation requirements related to the environmental impact assessment (or World Bank guidelines).

At the construction stage, when disruption management is necessary, the corporate objectives are to prevent or resolve conflicts that could impede finishing construction on time and on budget, and to strengthen local public support for the project by providing jobs and business opportunities and mitigating any adverse impacts.

During operations, engagement between corporations and stakeholders normally centers on the management of environmental and social impacts and on the monitoring, reporting, and auditing of sustainability performance. The specific corporate objectives for engaging stakeholders during operations are (1) to identify and address environmental and social issues and concerns; (2) to monitor, advise on, and verify audits of environmental and social performance; and (3) to facilitate the delivery of complementary programs of sustainable development. Gradually, the engagement with stakeholders at the operational stage is assuming the form, at various mines, of formally structured, ongoing oversight committees of stakeholders. An example is the Porgera Environmental Advisory Committee (PEAK) at a Placer Dome operation in Papua, New Guinea.

At closure, the objective is to prepare local communities for the transition to a postmine period and to implement programs of environmental rehabilitation and social adjustment. At some sites, it is becoming the practice to implement an ongoing postmine program of extended care for the environmental and social issues that persist after the mine is closed.

Stakeholder engagement occurs not only at the operational level but in some mining companies at the corporate level as well. The objectives of the corporation for engaging NGOs and other stakeholders at the corporate level are (1) to identify priority environmental and social objectives of a global nature; (2) to provide an early warning of issues of concern; and (3) to expand public and investor approval of the corporation. Since 1997, Placer Dome has held stakeholder roundtables at approximately 6-month intervals at various localities (Sydney, Australia; Vancouver, Canada; Washington, D.C.; and Denver, Colorado).

Meetings generally involve a dozen or more NGOs and half as many Placer Dome staff. The themes of the meetings have evolved from initially exploring the concept of sustainability, to developing indicators of performance, to identifying possibilities for monitoring progress and verifying reports, to developing requisite sustainability competencies, and finally to establishing structured approaches for monitoring progress, advising the corporation, and addressing specific issues. The experience of Placer Dome is that this process has accelerated the corporation's decisions on significant issues and concerns to stakeholders, has deepened mutual understanding and trust, and has provided a basis for engagement and possibly collaboration on a broadening range of global and site-specific issues.

CONCLUSIONS

It has become apparent in Placer Dome's relationships with NGOs that they are willing to work over the long term with the corporation only if they believe that they can truly affect the corporate agenda. In Placer Dome's experience, NGOs are affecting the corporation, both at mines and at the global strategic level. Placer Dome perceives significant value to be available from maintaining strong relationships with NGOs, and looks forward to eventually operating in partnership with NGOs in sustainability programs in the field. Experimental efforts in that direction are already underway.

Ultimately, NGOs will work in partnership with corporations only if they are certain that there is no risk of association with environmental damage, human rights abuse, labor exploitation, community oppression, or the erosion of national sovereignty. It is likely that eventually the global certification of the sustainability performance of mines will be required to provide this certainty. Placer Dome is exploring the concept of global certification with leading NGOs. For such a system to be implemented, however, will require the cooperation of a significant number of leading mining companies.

In general, the mining industry is consolidating a sectoral perspective on the principles of sustainable development, the procedures of engaging stakeholders, the approaches to monitoring and reporting on environmental and social performance, and the possibility of global certification of issues. Some mining companies support the UN's Global Compact with Business, which commits corporations to work with UN agencies and NGOs in advancing core environmental, human rights, and labor principles. A number of mining companies are members of the International Council on Metals and the Environment (ICME), which has recently engaged with a number of NGOs and multilateral institutions–notably the World Bank–to expand its charter to embrace the principles of sustainable development. Many mining companies are sponsors of the Mining, Minerals and Sustainable Development (MMSD) program of research and engagement with stakeholders under the Global Mining Initiative (GMI).

The trend is clear. The shape of the future is emerging. Stakeholder engagement, including dialogue and ultimately partnership with NGOs, once a pathfinding activity, is gradually becoming a routine practice of mining companies. The gulf between the corporate world and civil society, between mining companies and NGOs, is slowly closing, as we discover that our shared interests are more important than our differences.

2B or Not 2B? E-Commerce in Mining

Donald E. Ranta*

ABSTRACT

Many portions of the mining industry are under intense pressure to reduce costs while commodity prices languish. In response, global Internet-based marketplaces to deliver business-to-business (B2B) e-commerce solutions are being established to reduce costs and improve efficiencies across the mining industry supply chain as well as with customers. Over time these marketplaces will likely evolve to create an integrated supply chain solution, layer in value-added services, and develop mining community portals with industry content.

Significant opportunities to help reduce costs and create value are available in mining e-procurement as well as in on-line product sales. The supply chain for the industry is highly fragmented and many mining companies have numerous inefficiencies in procurement. An e-procurement marketplace (or "exchange") can immediately address these inefficiencies and offer numerous benefits to buyers and suppliers. These benefits include improved sourcing options, inventory cost reduction, transaction cost reduction, access to value-added services, lower customer and supplier acquisition costs, improved reach, and improved information flows.

GLOBAL B2B MARKET

The accelerating changes in business interactions through the Internet are more dramatic and more rapid than any other technological innovation since the industrial revolution. This interaction is termed "business-to-business (B2B) e-commerce," which was coined approximately 2 years ago to describe Internet-based commerce between companies. B2B marketplaces built for e-commerce are expected to create profound changes in the global economy by driving down costs and improving efficiencies across the supply chain throughout all industrial sectors, thereby creating value that will translate to greater competitiveness and higher market ratings for the winning players (Sood et al. 1999; Roskill and Cocroft 2000).

Commerce between businesses makes up more than 70% of the global economy. The growth of B2B e-commerce is projected to substantially eclipse that witnessed in business-to-consumer (B2C) e-commerce over the last few years. B2C players such as Amazon have revolutionized their industries and have earned multibillion-dollar market capitalizations in an unprecedented amount of time. According to Forbes.com (2000), more than 1,000 B2B marketplaces have been announced during the past year or so, and some are expected to enjoy

* NRX Global Corporation

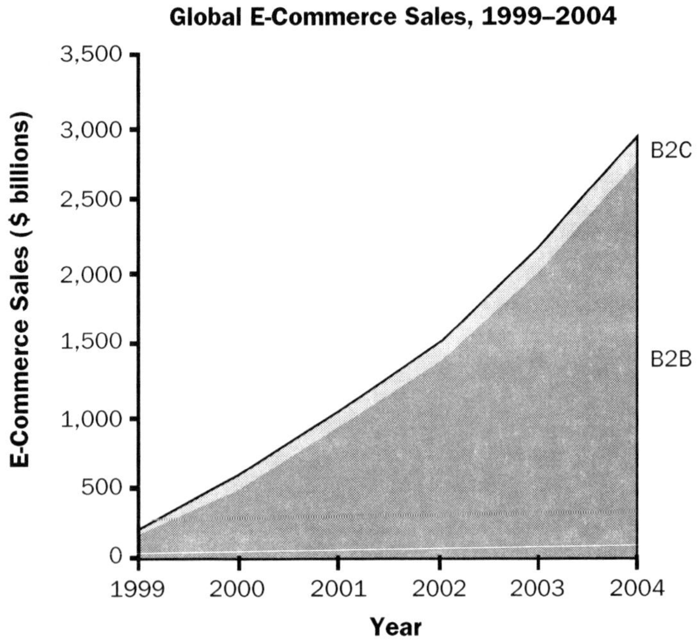

FIGURE 1 E-commerce market growth projection (*The Forrester Report* February 2000)

success similar to Amazon's. Total global e-commerce (B2B plus B2C) is projected to increase in sales from an actual $215 billion in 1999 to a projected $3 trillion in 2004 according to *The Forrester Report* (Kafka et al. February 2000). Of these totals, B2B e-commerce represents an increasing and overwhelming majority of the total e-commerce sales, with global sales increasing from 86% of the total in 1999 to an expected 93% of the total ($2.7 trillion) in 2004 (Kafka et al. February 2000).

Figure 1 illustrates the expected growth of the global e-commerce market, showing the relative B2B and B2C proportions.

B2B MODELS

Internet-based marketplaces can slash costs by linking companies with their suppliers and customers; improving inventory management; automating fax and phone procurement processes; and providing inexpensive sales, marketing, and customer support channels.

Variations of the B2B model include:

- Individual companies moving buying and selling on-line to cut costs and speed processes (e.g., Dell, General Electric)
- Third-party exchanges of independent firms bringing together multiple buyers and suppliers to create genuine marketplaces (e.g., FreeMarkets)
- Major companies of an industry getting together to form consortium marketplaces (e.g., Minerals & Metals eMarketplace) and inviting suppliers to join

Tapping into the flow of information and commerce among companies will become a competitive advantage and a potential source of profit for the marketplace controlling the transactions. The greatest question is who will control the marketplace–the independent "dotcoms" or the incumbents (major companies). Many of the majors are either refusing to join the independent exchanges by forming industry joint ventures or consortia or demanding significant equity stakes in startups in return for their participation. According to *The Forrester Report* (Kafka et al. August 2000), consortia know how to wring out supply chain inefficiencies, but independents will be better at providing market access and ensuring neutrality, which is critical for full buyer–supplier participation. Generally, independents will also become operational more quickly and have the advantage of speed.

To form a marketplace, infrastructure (equipment that powers the Web) and a full suite of B2B software must be acquired or licensed for use. The software must accommodate catalogs, auctions, exchanges, content management, and value-added services. This software links buyers and suppliers in creating a marketplace with dynamic pricing for the procurement or selling of materials, goods, services, and products. Catalogs from a variety of suppliers in a given industry are gathered and placed on-line; these enhanced catalogs are dynamic, three-dimensional, and interactive. Auctions and reverse auctions (where a buyer awards business to the lowest bidder) provide a forum for dynamic trading (where there is no set price). Exchanges offer constant price adjustments as supply and demand ebb and flow, similar to a stock exchange (Schonfeld 2000).

VERTICAL AND HORIZONTAL MARKETPLACES

B2B e-commerce can be conducted in vertical (industry sector) or horizontal (multi-industry) marketplaces. Marketplaces include companies that sell to customers or that buy from suppliers and distributors over the Internet.

Vertical marketplaces focus on specific industries, typically for the buying and selling of "direct" (production-related) goods and services. Key success factors for vertical markets are deep domain (industry) knowledge and strong industry relationships, which are necessary to gain high buyer and supplier participation. In today's early stages, vertical exchanges typically focus on either procurement or product (e.g., commodity) sales, but not on both. Over time consolidation of vertical marketplaces is expected, and the survivors will likely offer both procurement and product sales. Verticals would include industries such as mining and natural resources, chemicals, and automotive, among others.

Horizontal marketplaces focus on key business functions by offering "indirect" goods and services that are used across a variety of industries, and thus are not industry specific. Typically, these goods and services are labeled indirect in that they are not critical directly to the operations of the purchaser. Examples of these would be costs of MRO (maintenance, repair, and operations), office supplies, travel, and shipping.

Figure 2 illustrates the market components and interrelationships of vertical and horizontal markets.

In the near future, vertical marketplaces that have deep industry expertise will partner with horizontal markets or other service providers. These providers of functional knowledge will work with the marketplaces to create multiple offerings that incorporate value-added services tailored to meet the needs of specific industries.

ELECTRONIC COMMERCE

The concept of full electronic commerce using the Internet will create a paradigm shift in how B2B transactions are conducted. The Internet will allow communication between organizations on a level that has not been possible to date without huge expense and effort. In the past, electronic commerce has been the domain of electronic data interchange (EDI). With the advent of the Internet, and of exchange technology in particular, it is possible to enable collaboration and communication between companies through integration with their existing business systems. It has now become possible to do this at a reasonable cost and level of effort, within a universally accepted standard.

EDI was an attempt to electronically link a company's back-end enterprise resource planning (ERP) system directly with a supplier or a customer, thereby reducing the cost and time

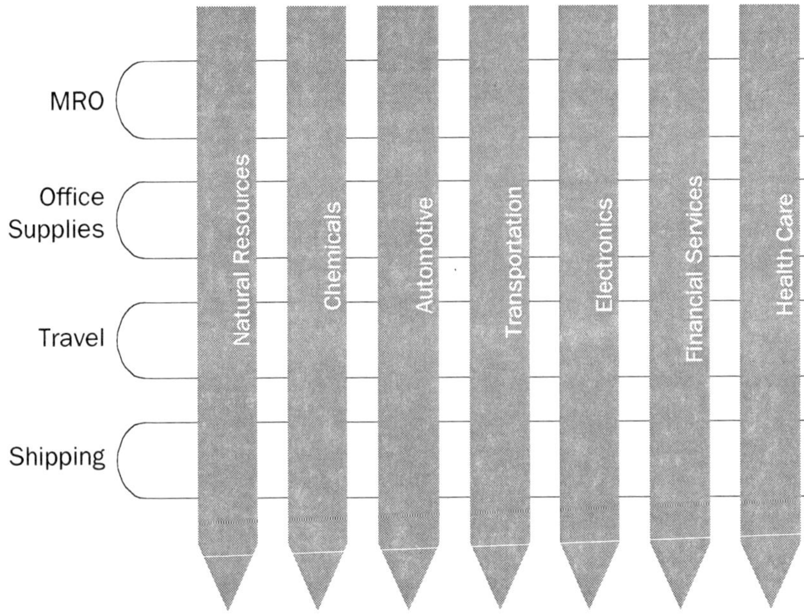

FIGURE 2 Market components

of transactions. Despite the exploding use of the Internet, EDI is still growing at a rate of nearly 15% per year (Sliwa 2000), clearly showing that this technology is perceived to create value to the buyer–supplier relationship. The greatest value comes through electronically integrating the business processes of the two parties, and the Internet can foster and enhance this relationship.

EDI is both expensive and cumbersome, which practically limits its use in automating large volume transactions between a single buyer and supplier. Automating each additional buyer–supplier relationship requires a new EDI implementation project. A marketplace changes this. In the traditional model, for every supplier a mine wants to EDI enable, a separate electronic mapping is done on both the mine side and the supplier side. This leads to multiple mappings on each side, as EDI networks are extended to multiple suppliers and mines as shown in Figure 3.

In contrast, the exchange model requires only a single electronic mapping for each party to the exchange (Figure 4). The exchange then channels multiple transactions through these mappings, making it cheaper and easier to maintain. The exchange acts as a router, through which it can transmit a company's e-commerce transactions to any other company it wishes. Immediate savings come through reducing the number of electronic mappings required, as well as from eliminating EDI charges. All former EDI transactions can now flow directly through the exchange. This dramatically reduces the amount of maintenance required by each party. It also means that where there is an existing relationship between buyer (the mine) and supplier, there are cost advantages to utilizing the exchange for this relationship. Where there is no EDI, it is now possible to create the same advantages EDI offers through the exchange at a reasonable cost. A supplier can create one mapping, then get the transaction flows to all of its mining customers that use the exchange. The same holds true for each mine and all its suppliers.

It should be noted that existing investments in EDI can be integrated into the exchange. The transactions conducted in the old EDI communication language need to be converted to extensible mark-up language (XML) so it can be interpreted by the exchange. XML is far more customizable than EDI, thereby allowing many more players to participate in an

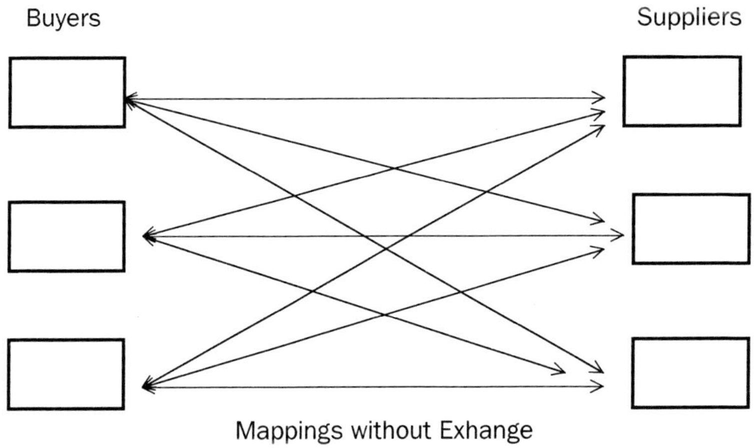

FIGURE 3 EDI configuration

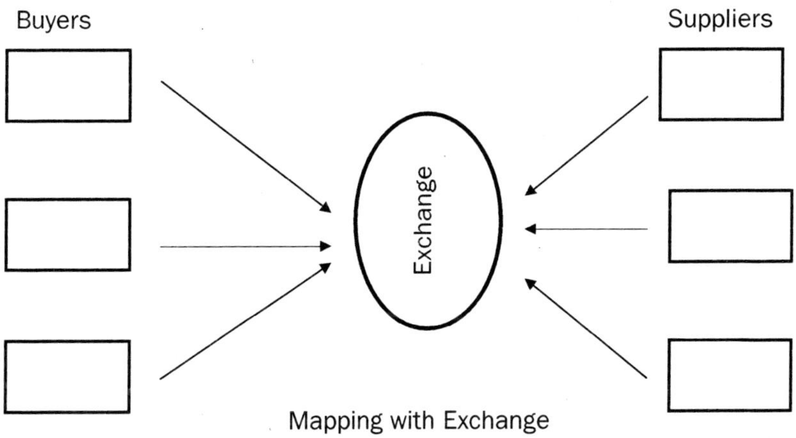

FIGURE 4 Marketplace (exchange) configuration

exchange at a significantly reduced cost. Similarly, transactions coming from the exchange will need to be converted to EDI before they are incorporated into an EDI-enabled company's internal systems. Several companies have developed software (known as "middleware") to accomplish this task.

To take the concept to its logical conclusion, integration of a company's back-office ERP system into the exchange allows the whole process to become seamless–beginning with the original request for purchase from the buyer through the marketplace to the supplier, and eventually back in through the e-payment transaction. Assuming the ERP system is electronically accessible from a remote site, once the purchase request is received and accepted, it will be reflected in the ERP system, thus triggering the payment cycle. As transactions flow through the exchange, each affected party can track the transaction status, either directly through the exchange software or by e-mail notification.

From this, it can be seen that the marketplace becomes the hub of a complete "virtual" e-business network that has been created around a particular buy/sell transaction. According to the Aberdeen Group (2000), " ... e-business will operate not as a single company but rather as a node on an integrated supply network, which leverages the Internet to exchange and process information in a manner that enables rapid response to customer requirements, using optimal mix of resources at the lowest possible cost."

PROCUREMENT AS A BEGINNING

The initial focus of e-marketplaces is the efficient matching of buyers and sellers, thus reducing transaction costs and easing trading relationships. Exchanges enable this through standard functionality such as catalog purchasing, contract purchasing, requests for proposals (RFPs)/requests for quotes (RFQs), auctions, and reverse auctions. The buyer needs only a standard Internet browser to interact with the exchange to search for, source, and procure goods. The resulting benefits include reduced transaction and material costs, improved sourcing options, reduced inventory, lower customer-supplier acquisition costs, expanded reach, and improved information flows.

But procurement is just the beginning of the value of B2B. According to some analysts, the term B2B should be discarded in favor of the general corporate use of the Web, which can be termed "corporate Internet services." Mining companies can use these expanded Internet services to change their way of doing business. This is where the exchange represents exciting new possibilities.

In addition to streamlining procurement, exchanges can create and capitalize on even greater opportunities. They can:

- Create true electronic linked transactions between buyer and supplier, thus cutting the transaction costs for both parties.
- Create an end-to-end transaction capability. This includes areas such as logistics, customs brokerage, insurance, and financial settlements.
- Collaborate on project management and engineering. By allowing all parties access to a single view of a project, whether it be engineering R&D or a new mine construction, there are tremendous timing and cost synergies that can be captured.
- Stay on the leading edge of technology and significantly reduce information technology (IT) costs through application hosting using an application service provider (ASP) model—essentially, software rental over the Internet.

BEYOND PROCUREMENT IN MINING

Mining is a large and attractive industry for B2B solutions. Significant opportunities to help reduce costs and create value are available in mining e-procurement. The supply chain for the industry is highly fragmented, consists of localized suppliers, and has poor information flows, large inventory costs, and limited vertical integration. Many mining companies also have numerous inefficiencies in procurement because of unconsolidated buying processes, maverick purchases, poor access to information, and relationship-driven supplier selection. An e-procurement marketplace can immediately address these inefficiencies and offer numerous benefits to buyers and suppliers.

E-procurement marketplaces and buyer-supplier relationships will act as catalysts to create an integrated supply chain solution with horizontal value-added services layered across the entire industry. Some of the key marketplace elements to a long-term solution, with benefits for both buyers and sellers, include:

- Dynamic trading forum—electronic trading, dynamic catalogs, RFPs and RFQs, on-line auctioning, and aggregated purchases
- Value-added services—electronic payment, logistics and shipping status, wireless solutions, risk management, certification and standardization, training and education, product life-cycle management, and collaborative design

- Supply chain integration–optimized inventory management, enhanced demand forecasting, information visibility, and buyer-supplier collaboration
- Community portals–creation of an industry network, industry content, and virtual office

These benefits will be achieved only through strong partnerships. No single buyer, supplier, or exchange can go it alone. Each marketplace must develop key relationships with a major B2B technology platform partner, sources of financing, key global mining buyers and suppliers, and industry content providers. These relationships are critical to maintaining the neutrality and building the scale (volume) essential for a successful exchange.

A strong management team, which has experience and depth in the mining industry, financing, technology implementation and integration, and B2B Internet e-commerce, is also necessary. To be able to compete, a group must put in place all the components necessary to ensure a strong B2B e-commerce business solution to the mining industry. Most importantly, an exchange must be able to clearly articulate the benefits it provides. Once this is accomplished, mining buyers and suppliers will recognize the inherent opportunities and be eager to join the exciting and productive B2B e-commerce environment.

REFERENCES

Aberdeen Group. April 2000. The e-Business marketplace: The future of competition, an executive white paper. Boston, Mass.: Aberdeen Group. 20 pp.

Forbes.com. August 2000. Best of the Web. B2B survey. August 24, 2 pp.

Kafka, S. J., B.D. Temkin, B. Doyle, T.O. Brown, and P. Martin. August 2000. The eMarketplace shakeout. *The Forrester Report*. Stamford, Conn.: Forrester Research. 20 pp.

Kafka, S.J., B.D. Temkin, M.R. Sanders, J. Sharrard, and T.O. Brown. February 2000. eMarketplaces boost B2B trade. *The Forrester Report*. Stamford, Conn.: Forrester Research. 18 pp.

Roskill, A.W., and C. Cocroft. January 2000. Business-to-business e-commerce. New York: Warburg Dillon Read, Global Equity Research. 113 pp.

Sanders, M.R., B.D. Temkins, T.O. Brown, and L.H. Hurd. July 2000. Unraveling eMarketplace deals. *The Forrester Report*. Stamford, Conn.: Forrester Research. 19 pp.

Schonfeld, E. June 2000. Corporations of the world, unite! You have nothing to lose but your supply chains! www.ecompany.com.

Sliwa, C. May 2000. EDI dinosaur lives–and grows. *Computerworld*. May 1, 2 pp.

Sood, R., J. Friedman, M. Parekh, R.G. Sherlund, L. Bahramipour, and T. Berquist. September 1999. *E-commerce/Internet, B2B: 2B or not 2B?* New York: Goldman Sachs. 94 pp.

The Internet: A Powerful Anti-Mining Tool in the Wrong Hands

R.J. McGregor*

ABSTRACT

The Internet has revolutionized worldwide communications. Any real or perceived mining incident or environmental spill, no matter how small or remote, can become a disaster of "biblical proportions" through the judicious use of the Internet by environmental and anti-mining groups. In addition, mining companies' own Web sites and unofficial message boards can provide plenty of additional ammunition to the anti-mining forces.

This paper explores how the Internet and readily available information are used by anti-mining groups active on the Web.

INTRODUCTION

The information available to anyone who has Internet access today is truly staggering. Checking theater tickets in London or the weather in Lusaka? Buying lapis lazuli from Santiago or opals from Lightning Ridge? No problem with the click of a mouse. By the same token, however, any real or perceived mining incident or environmental spill, even if small or remote, can turn into a major disaster quickly and easily over the Web. One only needs to carry out a superficial Web search to discover the multitude of environmental and anti-mining groups active on the Internet. With the ease of aperients, diverse groups from Australia to Canada can meet, talk, and compare notes and strategies. In addition, research on individual companies is readily available through mining companies' own Web sites, unofficial message boards, and various mining company databases.

COMPANY WEB PAGES

Most mining companies have their own Web sites to better communicate with their shareholders and other interested parties. The sites usually contain information on all of the company's projects and activities, along with all the latest press releases and copies of the main documents, such as annual or quarterly reports, required to be produced by a publicly traded

* Vista Gold Corporation

company. Annual reports also list the names and ages of all senior personnel. All this information is readily available through other sources. However, the ease and speed with which this information can be accessed by anyone, including anti-mining groups, through the Web has changed the landscape. Pictures posted on company Web sites to inform shareholders of activities in remote areas, which show, for example, drill roads cut into a mountainside, can appear out of context in a myriad different forms such as anti-mining posters. Along the same lines, pictures of steam rising from drying kilns or power plants have been used by environmental groups, in the past, to demonstrate industrial "pollution."

Most company sites are made even more accessible by their inclusion in various mine databases. This makes doing research on mining companies and their worldwide activities extremely easy for anybody who cares to do it.

MESSAGE BOARDS

Another interesting source of unedited, unofficial, and often uninformed information about a mining company is the message boards that spring up, sponsored by companies such as Yahoo. For the most part, these message boards are relatively harmless forums for interested investors and others to exchange views about the company. But because they are uncensored and unedited, they can be used to spread rumors and misinformation very effectively. There is no accountability because every posting is done anonymously using a Web name. Because of the anonymity, rumors about ownership disputes, environmental issues, and globalization issues, among others can be spread in an attempt to scare shareholders into selling stock and discredit the company.

Disgruntled employees, both past and present, can also use this forum to create mischief. Companies have different philosophies as to how to interface with these message boards. Most just ignore them, but sometimes it is difficult, particularly when there is false, misleading, and objectionable postings about the company or its officers. The best advice is not to engage in the dialogue on the message board, however tempting, as the rules for public disclosure forbid you from selectively disclosing relevant information. If rumors begin to take hold and start to seriously affect share prices, the best defense is putting out a press release; if the postings are personally insulting, develop a thick skin. If you are an employee, do not use your company's message board.

ANTI-MINING GROUPS

The Internet is littered with environmental groups that come in all shapes, sizes, and locations. Among the more active is the Mineral Policy Center, based in Washington, D.C. According to Mineral Policy Center President Steven D'Esposito in a statement to the Subcommittee on Energy and Mineral Resources on August 3, 1999, the "Mineral Policy Center is an environmental organization dedicated to protecting the environment and communities from the adverse impacts of mineral development, and cleaning up pollution from past mining." He went on to say that the Mineral Policy Center believes in responsible mining that can and does occur on our public lands, but then he defined responsible mining in such a way that it precludes any mine currently operating in the United States or anywhere else.

Other sites like Forests.org put out sweeping statements such as "Gold Mining Wreaks Havoc for Indigenous Communities and Eco-Systems Worldwide." Still other organizations such as Greengrants.org show the basic arrogance of extreme environmental movements worldwide with the first line in the Overview on their Web page, which states that "the first assumption of the Global Greengrants Fund is that we cannot save just part of the earth, we must save it all." This of course presupposes that the world needs saving and that the members of this organization are the people to do it. Many other extreme and often humorous examples exist, which will be expanded on in the oral presentation.

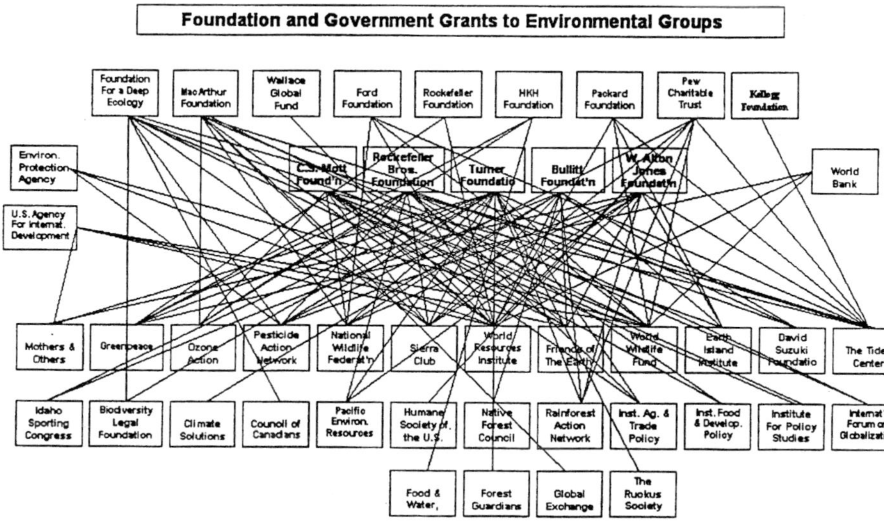

FIGURE 1 Foundation and government grants to environmental groups

An evening surfing the Web and reading the exaggerations, myths, and lies that are put forward as learned opinion and facts can be truly depressing, but take heart, for we are not alone.

THE WISE USE MOVEMENT

In the 1980s, a new grass-roots activist movement began to surface as a counterbalance to environmental extremism. Called the Wise Use or Common Sense Movement, it poses a significant threat to the environmental extremists, judging from some of the postings I have read. I believe the reason for this fear is that the Wise Use Movement tends to offer a more conservative environmentalism that acknowledges the importance of environmental protection but within the context that human beings have rights too, including the right to own private property and the right to life, liberty, and the pursuit of happiness! Many environmental groups cry foul, because for many years, the phrase "saving the environment" embodied the moral high ground. Now, however, equally powerful catch phrases such as "wise use" and "common sense" have emerged. Because these concepts are equally difficult to argue against, the extreme environmental groups attack the Wise Use Movement by "exposing" their corporate sponsorship. And while indeed there are both corporations and individuals contributing to the Wise Use Movement, the environmental groups carefully omit the fact that most of their funding also comes from corporate sponsorship. In addition, they fail to disclose how much money they get from foundations and government grants. Figure 1 shows some of the funding channels from foundations and government agencies to environmental groups. On the figure, I see no Wise Use organizations on the receiving end of this large amount of sponsorship.

The war against mining continues to be waged on the Internet by the Central Banks and by many public and private sector organizations around the world. We have a long way to go before the pendulum swings back into balance with respect to natural resource industries worldwide, but you should be aware of the debate so that you can participate and protect your futures.

. . . .

World Bank Group Policies and Guidelines (Did Someone Move the Goalposts?)

B.A. Filas[*] and D.W. Fohlen[†]

ABSTRACT

The International Finance Corporation (IFC), the private investment arm of the World Bank Group (WBG), requires all IFC-financed projects to comply with its environmental and social policies and guidelines. IFC's review process involves an environmental and social assessment and the implementation of an Environmental Action Plan (EAP) that will set forth the necessary actions to bring the operation into compliance. This paper provides an overview of the environmental and social policies and guidelines that apply to mining projects and examines how WBG/IFC policies and guidelines are applied to mining projects in developing countries. In addition, it includes a review of the increased emphasis IFC is giving to social issues over the past several years based on the existing WBG policies.

IFC's process for the environmental and social review of proposed investments is briefly discussed. The paper focuses on how IFC emphasis has shifted from the traditional environmental and impact mitigation analyses of a few years ago to the integrated social and environmental awareness, action (EAP), and implementation programs (social plans) required today. Today, IFC's focus is to promote sustainable development by investing in projects that directly benefit local communities and the environment. Examples of these changes will be highlighted by a review of recent investments located in West Africa. The goalposts have not moved, but the rules of the game have become tougher.

INTRODUCTION (THE NAME OF THE GAME)

The World Bank Group (WBG) is a development bank that provides financial assistance to the private sector through loans, equity investments, and other financial instruments through its private investment branch, the International Finance Corporation (IFC). Mining companies often look to IFC for financing when the investment risk in a developing country may exceed the lending policies of the commercial institutions. As a multilateral financial institution (MFI), IFC has strict policies and guidelines for promoting sustainable development

* Vice President, Mining and Environment, Knight Piésold Services, Inc.

† Senior Environmental Specialist, International Finance Corporation

through thoughtful social and environmental program development. When a company approaches the IFC for financial assistance, it must be fully committed to undertake the environmental and social responsibilities that the WBG requires.

Although none of the WBG/IFC environmental and social policies and guidelines will seem particularly onerous to companies committed to the principles of sustainable development, they often far exceed what the local regulatory authorities in developing countries would otherwise require. As such, these policies and guidelines very often set the standard for environmental and social management in developing countries around the world. Where laws and regulations do not exist, the internationally recognized standards for environmental compliance are often the WBG guidelines. In addition WBG social and environmental policies drive a best practice approach in many developing countries. The WBG track record and lessons learned from various projects worldwide help to provide to investors a pragmatic and value-added approach on environmental matters. Regardless of WBG involvement, the policies still carry over to other projects not financed by WBG.

THE GLOBAL MINING DEPARTMENT (INTRODUCING THE PLAYERS)

Responsible use of mineral resources gives developing countries a considerable number of opportunities for poverty reduction and economic development. Countries, communities, and companies often face tough questions about opportunities and risks as they develop environmentally and socially responsible programs for mineral resource development. The WBG recognizes the potential embedded in a country's mineral sector to significantly influence regional and national economic and poverty profiles. To improve the effectiveness and reach of its activities in the mining sector, the WBG has combined International Bank for Reconstruction and Development/International Development Association (IBRD/IDA) and IFC operations in the mining sector into a single Global Mining Department. The new department aims to bring the work on the public-sector policy side and the private-sector investment support side closer together, to achieve synergies and to provide a simpler and more direct contact point for government, private-sector, and civil-society clients.

The Global Mining Department is actively promoting responsible mining, mainly in developing countries with little or no private-sector investments in natural resources. This recently created department now offers services to governments, the private sector, and to civil society. Its vision is a mining sector that, by attracting private investments, creates a foundation for economic and social well being and environmental responsibility. Having weighed the pros and cons, the WBG will continue to actively promote the mining sector in developing countries. By providing IFC financing, the WBG supports environmentally and socially sustainable private sector investments in West Africa and other developing countries.

FINANCING MINES IN FRONTIER COUNTRIES (THE PLAYING FIELD)

Mining projects in West Africa present a unique challenge to project proponents and also to their investors. Nearly all West African countries are underdeveloped and most are heavily reliant on foreign assistance to support the existing government structure and its day-to-day operations. Government structures may or may not be well defined, and regulatory programs affecting mining activities are usually also ill-defined. The authors' experience has been that government agencies in most developing countries have few laws and regulations on the books that are applicable to mining activities and environmental concerns. This inexperience in regulating the mining sector is usually evidenced by regulatory programs that have little or no funding to implement and enforce the laws, insufficient or nonexistent administrative rules, and government authorities who have no experience in mining or environmental projects.

The economic importance of mining as an industry in West African countries varies–perhaps because of the significant capital investment that is required for development and the risky political environment into which those investments must be made. Some countries, such as Ghana, Mali, Sierra Leone, Senegal, and Guinea are comparatively well endowed with mineral resources; others have little influence from the mining industry. In nearly all cases, mine development is an expensive proposition, because most West African countries have under- or undeveloped infrastructure systems that require the mine developer to supply its own project needs with the needed infrastructure to support its operations.

Local communities in rural West Africa typically have no suitable infrastructure. Electricity, telecommunications, potable water, and efficient sewage systems are luxuries that most households do not enjoy. Education opportunities and access to quality health care are rare. Most of the population does not have access to transportation. These conditions leave the mining company faced with a typically undereducated and unskilled population available to serve its employment needs, living in an environment where diseases like AIDS and malaria may significantly affect longevity of the workforce. The spread of AIDS may even be exacerbated by mine development because of the influx of new people to the area seeking employment or providing supplies to the mine and its workers.

Although these conditions make mine development a particular challenge to the mining company, they present a unique opportunity for an MFI that is looking to invest in financially viable projects that will benefit the host country and its people in a sustainable way. The importance of community development programs in West Africa and other developing regions around the world is a growing emphasis. The WBG's emphasis on promoting economic development opportunities in sub-Saharan Africa leads to a win-win opportunity for mining companies seeking financial assistance in the region.

ENVIRONMENTAL AND SOCIAL REQUIREMENTS (THE RULES OF THE GAME)

IFC's environmental and social review procedure is intended to help the mining company to focus on what will be important to assess on the environmental and social aspects of a project. It establishes the ground rules and advises a company up front of what environmental and social obligations will be expected of it if it intends to gain IFC financing. This procedure has been developed both for IFC staff and the mining company at the appraisal stage in order to make a comprehensive assessment of the project and what issues will need to be addressed in order to meet WBG requirements.

During the initial project evaluation and screening process, IFC must make a determination of the category under which the project will be evaluated. The IFC will typically assign projects into one of three categories depending on their degree of impact:

- Category A—A project is classified as Category A if it will result in significant impacts to the environment or social atmosphere. Proponents of Category A projects must retain independent experts to carry out the Environmental Assessment (EA). Where the project is particularly risky or sensitive, it may also be advisable to retain an independent panel to advise on key issues or aspects of the EA process.
- Category B—Projects resulting in less significant environmental and social impacts than Category A can be classified as Category B. Although variable from project to project, the scope of the EA may be narrower than that for Category A, but the basic elements to prevent, minimize, mitigate, or compensate for adverse impacts remain.
- Category C—When project implementation will have little or no impact on environmental or social values, it is classified as Category C. Once classified as Category C, no further environmental or social analysis is required.

Nearly all mining projects will be classified as Category A, requiring the development of rigorous environmental and social documentation to quantify the effect that the project will have on the surrounding environment. Occasionally a small project or an expansion of an existing mine will be classified as Category B if the environmental and social impacts are adequately documented or managed, or both. A new mining project, or the reopening or retrofitting of an existing major operation, will usually be classified as Category A.

IFC has developed "guidance notes" to focus the project proponent on the important issues for the project and to outline what will be required for the project. This process is important in advising the mining company of what will be expected of it when approaching the IFC before it proceeds. These notes include a checklist of potential issues and they recommend the contents that should be addressed in the environmental impact assessment, environmental action plan, environmental audit, major hazard assessment reports, and public consultation and disclosure program.

IFC will review a project for conformance with its established WBG policies and guidelines. These policies are typically presented as WBG Operational Policies (OP), WBG Operational Directives (OD) or IFC Safeguard Policies. They define the policy and processing procedures for projects that may affect a particular discipline. Industry specific guidelines (such as base metal and iron mining) have also been developed that establish what information and standards the WBG/IFC recommends to a project proponent. Guidelines are usually specific but flexible for a given industry, as opposed to policies that are more comprehensive but may not be relevant to every project.

THE ASSESSMENT PROCESS (PLAYING THE GAME)

To be eligible for IFC financing, the mining company must clearly demonstrate that the project is technically sound and profitable for the investors, a benefit to the host country economy and local communities, and compliant with stringent environmental and social policies and guidelines. IFC is likely to review the environmental track record of the mining company and also advise on the implementation of sound environmental and social management programs. Proponents must provide sufficient economic, financial, technical, legal, environmental, and social information so as to allow the IFC to determine if the project is financially viable as well as socially and environmentally defensible.

The amount of debt and equity financing that IFC will provide for any single project is 25% of total estimated project costs. This is to ensure the participation of investors and lenders from the private sector. It may provide up to 35% of the equity capital for a project as long as it is never the largest shareholder. As a substantial financier of mineral projects in developing countries, IFC is able to act as a catalyst for the investment of additional private sector debt and equity funds in the projects it finances. It can often mobilize financing directly by syndicating loans with international commercial banks and underwriting investment funds and corporate securities issues. It can also handle private placements of securities and various risk insurance instruments.

Early in the process of considering a project for financing, IFC environmental and social specialists will visit the project to gain an appreciation for the local site conditions, social issues, and regulatory purview. This visit allows IFC staff to maintain a proper perspective on the project when they review the feasibility study and the environmental and social impact assessments. They will be particularly interested in what measures the proponent has in place to inform the public on project activities and will make recommendations as needed on ways to better enhance and document those communications.

Mining companies are well advised to participate with IFC on these visits and if possible have their independent environmental expert participate as well. This will help to establish a clear understanding of the issues that must be addressed in the environmental and social analysis and also to seal the working relationship among the project sponsor's specialty teams and IFC staff.

The Operating Policy 4.01 (OP 4.01) defines the IFC policy for EA and comprises the following fundamental components:

- The Public Consultation and Disclosure Plan (PCDP)
- The Environmental Impact Assessment (EIA)
- The Social Impact Assessment (SIA)
- The Environmental Action Plan (EAP)

The first step in the process is the preparation of a Public Consultation and Disclosure Plan (PCDP). This plan defines how the company will collect and disseminate information with the people who are interested in and affected by the project. It establishes the basis for how the company will conduct its business with the community, and if implemented as presented, will assure transparency in project activities as they may affect the health, safety, and environmental condition of the people. The mining company representatives should prepare the PCDP itself, but with support from their consultants. The purpose is to reflect accurately what the sponsor is prepared to do in the public awareness process, during consultation with stakeholders and for the disclosure of appropriate reports. IFC requires that the PCDP be on file before they will invest significant time into any of the other aspects of the project.

The EA process required by IFC is analogous to the environmental assessment process required by most established regulatory programs. It consists of conducting detailed investigations to characterize the baseline conditions of the natural environment (air, water, and land values). The project development plans and alternatives are then overlaid onto the various baseline resource layers, which allows for detailed analyses of the project impacts on each receiving media within the regulatory framework that is applicable to the project. As impacts are identified, mitigation measures are applied to limit the severity of those impacts to levels that are acceptable to the regulatory communities, interested stakeholders, and the public.

The SIA required by IFC may tend to deviate from the conventional baselines and impact analyses that mining companies may be accustomed to, primarily because most IFC investments occur in developing countries where the social issues are unique. Mining projects may affect indigenous peoples, or require that individuals, families, or even whole communities be relocated to accommodate the mine development and operations. In this context, the socioeconomic baseline characteristics must present an in-depth characterization of not only the local economic and social structures that are normally included in a socioeconomic evaluation, but it must also consider their unique cultural heritage, social organizations, religious beliefs, and other distinctive behaviors.

Especially when resettlement issues are involved, social baselines and impact assessments must clearly inventory who will be affected by project development, how they will be affected, and what will be done to ensure that their situation after project development is as good as, or better than, their situation before project implementation. IFC requires very specific documentation of this interactive process with affected peoples, and a clear presentation of how affected peoples will maintain equivalence.

The Environmental Action Plan (EAP) defines the mitigation, management, monitoring, and institutional measures to be undertaken to eliminate adverse environmental and social effects, offset them, or reduce them to acceptable levels. It defines the specific actions that the project sponsor will undertake to implement these measures and to document the environmental and social performance of the mining operation. In addition to the specific action items, project sponsors are also required to present a schedule and budget for the implementation of the actions identified in the EAP.

Where appropriate, the EAP will be complemented by individual plans prepared for specific activities or aspects of the project. Most projects will require the development of one or more component plans. These plans define specific procedures, actions, and monitoring that are needed in order to accomplish certain tasks that will be required by the project. The component plans will vary from project to project, but some of the more common component plans required for mining projects in developing countries are:

- A Waste Management Plan that defines the specific actions and engineered systems for the handling, storage, and disposal of wastes that are generated by the project activities.
- An Emergency Response and Contingency Plan that presents responses to emergency situations that may occur at any time and identifies specific scenarios and actions that a first responder should take to provide for the safety of himself and others, and to limit the extent of effect on the environment.
- A Community Development Plan that presents the actions that will be undertaken to further community development opportunities and to alleviate pressures that may be experienced by the influx associated with the mine employment opportunities. This plan is fundamentally hinged on an effective public consultation program to identify program initiatives, and will include stakeholders in the decision-making and financial dispensation process.
- A Resettlement Action Plan that presents the specific details for relocating individuals who will be displaced by project development. Disruption of community structures, social networks, kin groups, and loss of productive assets and income sources can cause severe long-term hardship, impoverishment, and environmental damage unless appropriate measures are carefully planned and carried out.
- An Indigenous Peoples Development Plan that addresses the effect that project development will have on indigenous people, tribal groups or ethnic minorities with a distinct social or cultural identity. The plan must define the specific ways in which the mining company intends to involve these groups in the planning and development process and to mitigate adverse impacts that may be caused to them by project implementation.
- A Dam Safety Review by an independent panel of experts is required for all impounding structures exceeding 15 meters in height. IFC has the discretion to require an independent review for smaller dams when the design is particularly complex or if it will contain toxic materials. The independent reviews will occur throughout the investigation, design, construction, and startup of the facility. IFC requires that impounding structures be supervised by qualified companies during construction and that the quality assurance protocols and periodic safety inspection schedules are followed.
- A Mine Reclamation and Closure Plan describing how the project proponent will reclaim the site disturbances to a condition that is compatible with the surrounding land form, ecology, and aesthetic character. The plan must include an estimate of the cost to close and reclaim the mine site and provision for cost accrual into an escrow account (or equal) that is administered by an independent third party.

When the project environmental assessment is complete, the project documents are submitted to IFC for review and approval. IFC staff will review the documents for conformance with its environmental and social policies and guidelines, as well as for its robustness with respect to project financials. There is usually a period of questions and answers between the IFC and the project sponsor and its EA consultants to ensure that the documents provided fully conform to IFC policy. When IFC staff is satisfied that the policies and guidelines are adequately addressed, and that relevant programs are adequately designed to manage environmental and social effects, the project is made available for public disclosure.

Public disclosure consists of posting the project documents in the WBG InfoShop and making them available to the public in the locality and region of project influence. The InfoShop is the disclosure house for all WBG documents. Interested parties and stakeholders can access the InfoShop either directly or via the Internet. Disclosure documents can be ordered in hard copy or downloaded from the Internet. In-country documents are typically made available to the public and local communities directly from the project proponent. Once the documents are available to the public, the project sponsor must make the local communities and interested stakeholders aware of their availability. Notification is often made through local and regional news media, but it must also be made in a way that is understandable to the local public.

The project documents and information must be available to the interested public for at least 60 days before the IFC investment decision. In rural West Africa, public disclosure means verbal communication at public meetings because many members of the general public are illiterate. In some instances, local dialects may be unique to the region and the language of the project documents may not be readable or understandable by local interested parties. In these instances, it is the responsibility of the project proponent to articulate the project plans and environmental and social evaluations to the public.

The IFC Board of Directors has ultimate authority to approve or deny financing based largely on the recommendations of the IFC staff. Before going to board approval, IFC staff will recommend specific conditions of disbursement and environmental and social covenants in instances where additional information is needed from the project sponsor. Such conditions require that the outstanding information be provided before the investment money is issued. All IFC investments require ongoing project supervision, monitoring, and reporting after the investment is made. The specific commitments for environmental and social controls, monitoring, and reporting will already have been defined and committed to in the EAP.

TRANSITION FROM ASSESSMENT TO ACTION (STRENGTHENING THE RULES)

From the company perspective, effective implementation of environmental and social programs begins with sound company policies on environmental and social issues. It requires the commitment of the senior and executive management of the company to ensure that the programs are adequately implemented and financed. Mining companies must be committed to a long-term involvement in the country they are working in and to the people who live there. The fundamental philosophy of the mining company must be to return something to the people of the countries in which it operates.

The key to sustainability is to develop good projects that take advantage of the mine profitability during times when the company can afford to invest in the community. The proponent should act as a corporate citizen in the country. It will maximize the positive developmental impacts of the project at the local, regional, and national levels, but at the same time focus the company investments in a direction that promotes self-sufficiency in the community, rather than dependency on the mine economy.

The investment of mining projects in the environment and the community is only a small step toward achieving sustainability in a developing country. The concept of sustainable development recognizes that it is not just the traditional measures of economic welfare that matter to a project; it recognizes that development is essential to satisfy human needs and to improve the quality of human life. Although one project will not fulfill these obligations on its own, every step is a significant contribution toward reaching the sustainability goal in the region.

The principles of sustainable development can be easily lost without the support of the people who are affected by the project. Transparency of health, safety, social, and environmental activities is critical to keeping the public confidence high in the project. Without an efficient tracking of what is actually happening at a site; the environmental performance of the operation; and the effect of those activities on the people, the communities, and the receiving environment; there is no profitable long-term investment in a West African country. Poor relationships with local communities and regulatory authorities can affect the day-to-day performance of the mine.

The new cornerstone of IFC's sustainable approach is the sound implementation of the EAP. The main step is an effective means of disseminating that information to the stakeholders. If not successful, the environmental management program is of limited value. The project sponsor, through all management tiers of the organization, must ensure that proper attention is given to assess the effectiveness of the environmental mitigation, control practices, record keeping, and public consultation programs. This requires monitoring, comparison of results with target levels or objectives, implementation of further protection or corrective actions if warranted, and, as appropriate, notification of the interested stakeholders of the conditions and occurrences at the site. These activities are required at all stages of project development, from construction through operations and closure.

OUTCOMES—THE NEW FOCUS (A WHOLE NEW BALL GAME)

It is usually easiest to point out changes in a procedure or process through the use of actual examples. The authors are familiar with two projects that were evaluated under analogous environmental and social settings in West Africa. The documentation for one project was prepared in 1997; the documentation for the other was prepared in 2000. Both were prepared to address the requirements for a Category A mining project and both were proposing to mine the same commodity using similar mining and processing methods. Salient observations between the two submittals are:

- **Environmental Assessment → Social Assessment:** There has been a shift in emphasis from environmental assessment to social assessment. This is not to belittle the need for adequate and accurate environmental assessment, nor is it to intimate that social assessment was not a part of the early EA process. It is simply to suggest that the relative weighting of the environmental aspects to the social aspects in an EA prepared for IFC consideration has shifted. This shift is believed to be the result of two circumstances. First, the art and science of the environmental assessment process are well understood and EA specialists are well versed in environmental impact analysis to the extent that adequate EAs are commonplace. Second, social impact assessments are not so straightforward. It is not so common to find a specialist who has expertise in the socioeconomic aspects, as they are unique to every region of the world. Qualified social development specialists with relevant expertise in a given region are very hard to come by. The uniqueness of the social considerations in West Africa is significant, and the issues can be quite sensitive.

  Adequate management of the social issues to be prepared up front is critical to a successful project in West Africa. Although they have always been a part of the IFC process, the SIA and the complementary component plans like Resettlement Action Plans, Indigenous Peoples Development Plans, and Community Development Plans are now much more common for mining projects than they were in the past. In addition to the plans themselves, the increased levels of detail being required in them are all spinoffs of this increased emphasis on social understanding.

- **Impacts and Mitigation → Actions and Implementation:** The emphasis has also shifted away from the traditional impact analysis in order to identify suitable mitigation measures that would reduce the level of significance of those impacts to acceptable levels. Again, this part of the process remains an important aspect of the analysis. IFC has simply expanded its expectations to require specific definition of the process by which those controls and mitigation measures are actually implemented. Nowadays, it is not enough to simply identify what the controls and measures are, the proponent must also clearly articulate how it intends to demonstrate that its controls and mitigation measures are performing according to their intended purpose.

  The specific requirements for the EAP to clearly define the plans to monitor and measure results are a direct result of this increased emphasis. The EAP must identify what, how and when specific activities will be done. It specifically identifies who is responsible to assure that each activity is accomplished and the chain of authority under which that responsible person functions within the organizational structure. It also defines how the results of the various activities will be analyzed; as well as how, when, and to whom the results will be reported.

- **Closure and Reclamation → Financial Assurance:** Closure and reclamation activities have always been the primary method for mitigating environmental impacts associated with mine development. Again, closure and reclamation activities remain critically important to the EA process and the importance of a well-thought-out closure plan is key to a successful EA. However, in line with the shift toward actions and outcomes, IFC is now asking for assurance that the cost of closure and reclamation activities is accounted for. The mining company is, as a matter of standard practice, now asked to set up ways to demonstrate that it is accruing sufficient liquid funds to address its reclamation obligations at the end of the project life. These funds are typically required to be accumulated and administered by a third party to ensure that the money is available at the end of the project life when reclamation cost obligations will be great and project cash flow will be diminished. Several surety mechanisms are available to mining companies, but the use of escrow accounts that can be contributed to throughout the operating life of the facility on a per-tonne processed basis is often considered.
- **Up-Front Plans and Designs → Monitoring and Reporting:** Plans and designs that integrate preventive engineering to mitigate adverse environmental effects are common in the design process. Many engineering designs that mitigate certain environmental performance characteristics will often include monitoring recommendations from the designer in order to demonstrate design performance. Although this process and its relative importance have not changed, mining companies have had to go beyond simply identifying what monitoring will be done. Now, the EAP requires that the "how, when, why, and by whom" questions be answered clearly. The current program not only calls for monitoring and reporting, but also for clear demonstration that periodic inspections, audits, and public disclosure occur to an adequate level to ensure transparency.

CONCLUSIONS (WINNING THE GAME)

It is probably a misnomer to suggest that the WBG emphasis on environmental and social programs has shifted. Perhaps it is more accurate to say that the emphasis has expanded, and expanded considerably. All the evaluations and analyses that were important in the 1990s will continue to be equally important into the twenty-first century. It is simply that the list of what must be addressed and at what level of detail has become longer and more detailed. The diversity in social issues among developing countries where the WBG is involved, and the differences in the levels of performance among project sponsors in all bank-financed disciplines, has changed the game. To ensure that any investment will produce the results predicted by the EA and SIA, and will also be a benefit to the local community and the country as a whole, the WBG has simply expanded the playing field and strengthened the rules of the game. It applies first to the IFC, its private investment arm, because of what has been learned from both the good and bad experiences.

Many studies demonstrate that the private sector can play a leading role in sustainable development by creating a foundation for economic and social well being and environmental responsibility. By enforcing further its policies and guidelines, the WBG is actively supporting developing countries, investors and junior mining companies to promote long term and sustainable development. This new approach complements what major mining companies are trying to achieve in their operations worldwide.

Requiring additional emphasis on social aspects and on performance-based implementation assures stakeholders that IFC-financed projects of all types will live up to all their environmental and social expectations. Rest assured that the goalposts are still in the same place, but the distance to those goalposts and the rules that must be adhered to in getting there have changed significantly. It is still the same game, there are just many more variables that must be addressed in order to win.

BIBLIOGRAPHY

Central Intelligence Agency. The World Factbook 2000. Web site: http://www.odci.gov/cia/publications/factbook/index.html

International Finance Corporation. Web site: http://www.ifc.org

SECTION 4

Training and Technology

. . . .

Technology Transfer to the Mining Industry

Vaikuntam I. Lakshmanan,[*] Dan A. Macki, and Ramamritham Sridhar

ABSTRACT

Historically, introducing new technology to the mining industry has been problematic. In general, the industry is reluctant to take risks with new technology because it is already heavily tied to risk, even using existing technology.

Risk in the orebody itself can be substantial. The company is dependent on drill hole patterns that are expected to define what is underground, and process test work is often based on drill cores and other samples, which occasionally produce surprises. And even well-proven processing techniques can blindside the producer, resulting in embarrassing and costly difficulties.

But if the industry is to grow in knowledge and become more efficient, new ideas must be introduced. How to introduce new technology is the essence of this paper.

PEOPLE SKILLS

It has been said that technical problems are relatively easy to solve; it's the people problems that are more difficult. To approach transfer of new technology, it is absolutely necessary to have a strong leader and to create the right team.

CONVENTIONAL TECHNOLOGY RISK

Introducing new technology compounds the situation. This is particularly true in the mining industry, which is already burdened by the risks inherent in identifying and exploiting orebodies. One has the impression that the last thing it needs is to compound those risks by trying new processes or process modifications. The industry is sometimes blindsided by unexpected problems created by the ore itself or the machinery to process it, occasionally with disastrous results.

A case in point was the Gortdrum Copper Mine in Ireland. Halfway through mining the orebody, mercury was discovered in sufficient quantities to cause smelters to reject the concentrates, turning back whole shiploads. A mercury retort plant had to be implemented quickly, which had its own difficulties because of high arsenic levels. Extensive piloting

* Process Research ORTECH, Inc.

† Process Research Management, Inc.

solved the problems, but when startup began, unexpected equipment problems emerged, making operations difficult in the beginning. Once solved, the mercury plant became a major profit center for the company.

The Baya Loma heap leach solvent extraction–extrowinning technology (SX–EW) plant in Chile recently experienced problems with unexpected nitrates and chlorides in the ore. They were working through solving the problems at press time.

NEW TECHNOLOGY IS INEVITABLE

To say that mining people are gun-shy of new technology is an understatement. At this writing, however, the industry is under intense pressure from low metals prices and huge environmental concerns. Paradigm shifts in processing methods are inevitable and the industry desperately needs to embrace new challenges.

Historically, new technology is adopted slowly by the industry, but when one is well ensconced, it is welcomed by the whole spectrum of operators. Some examples are autogenous and semiautogenous grinding, column flotation, and application of SX–EW in copper recovery. Most have taken more than 10 years to get accepted. Most new technology is developed in laboratories like Process Research ORTECH, Inc., after which a commercial operation must be tried, preferably in a small plant. Usually a junior or maverick outfit is the first to take the plunge, often with pockets that aren't deep enough to carry through final development, and unless quick success results, the larger companies will not take on the process. A "show me" attitude prevails. The introduction of SX–EW at the Bluebird Mine in Arizona by ranchers in the 1960s is one exception, where the gamble resulted in outstanding success, transforming the industry almost overnight. At the other end of the spectrum, plants like the Coloso ammonia leach plant in Chile turned out to be a $250 million debacle and now sits like a monument, unused.

Approaching New Technology Transfer

Given that the transfer of new technology is inevitable, how do we approach introducing it to industry?

Certainly doing one's technical homework is a must. That means having work reviewed by one's peers, something that can be difficult if the technology has been under a cloud of secrecy or is under patent application. Many industries require intense secrecy, but that simply does not apply to the mining industry. Openness is the best policy. The industry is so skeptical of anything new that fearing that someone or some organization will take one's idea and run with it is almost ludicrous. The recommendation, therefore, is to submit one's ideas to carefully selected reviewers and to invite criticism. This can also be done by presenting papers to narrow or wide audiences, depending on what can be learned.

BREAKING THE FINANCIAL MOLD

A great difficulty in introducing new technology to the mining industry is the barriers that face those trying to get their new ideas to the marketplace. If a good idea has been developed at the pilot scale, it may be relatively easy to get the end user on board, even enthusiastic, about the method or equipment proposed. Problems arise when financing is being raised. Bankers or financial institutions want assurances to minimize their risk, so they turn to engineering companies for opinions. Then, because of increasing liability pressures, the idea is stopped in its tracks. Engineering companies recommend and sell technological expertise, but seldom in areas outside their scope of current practice. It is surprising how often the wrong technology is put into operation, simply because of entrenchment in current usage.

This user-banker-engineer cycle needs to be broken if the proponents of new technology are to succeed. The answer lies in demonstrating in a forceful manner that the risks in new technology can, indeed, be less than the risk in *not* using it. That means educating a broad spectrum in the benefits of the technology, not just the user, but the bankers, financial institutions, and engineers.

PEOPLE PROBLEMS

Once the technology has been satisfactorily scrutinized, introducing it to an application team is possible. People, being what they are, provide the biggest challenge, because the possibility of being blindsided by alternative agendas often emerges.

CULTURAL CLASHES

A large process plant was being built for a major metals producer in Canada, using technology that was unfamiliar to the company operators. A German manufacturer that was accustomed to doing turnkey plants throughout the world was providing basic engineering and supplying process equipment with project management through its Canadian office. Technology transfer would come through the design, construction, and commissioning steps. A Canadian consulting firm would provide the civil design and take responsibility for structural and other related design. The role of the Canadian project manager was to coordinate the German team with the Canadian engineers and to furnish administrative support and some drafting personnel recruited from the local labor force.

At the outset, there was a work ethic clash. The German team was accustomed to working alone on the entire project, using subcontractors where needed. Because they were used to coordinating through personal contact, not formal reporting, their reporting skills left much to be desired. They had few reporting systems, and therefore had to adapt to systems imposed on them by the Canadian project manager. They were used to making ad hoc changes in design without reporting to anyone. To complicate the issue, the owner had its own team that was imposing detailed changes to design and equipment supply, especially where Canadian or American equipment was being substituted for German equipment. The Canadian consultants, who were expected to stamp the drawings, were presented with some design assumptions that simply could not be applied because of local codes and site conditions. As a result, the Canadian consulting firm was often faced with design changes that simply appeared, and change orders mounted. From their point of view, it was a license to print money. Canadian engineers, who were subcontracted to the German team, also had their frustrations, because they were not allowed to provide any creative input to the process stream, yet they were expected to stamp the drawings and take responsibility for them. That brought on cultural clashes that resembled *Hogan's Heroes*, although it was only funny in hindsight. At the commissioning stage there were great arguments about North American equipment that had been substituted for German equipment and vice versa. The project startup did go smoothly, a tribute to the German company. The project cost, however, which had been estimated at $85 million, came in at $190 million.

What should have been a simple transfer of technology became a nightmare. The fundamental errors made in this case were: (1) there was no clear, designated leader; (2) there was no attempt made to consider the differences in work ethic; and (3) if a local consulting firm and local engineers are to be used, they must be involved in the project basic design and setup.

TEAM APPROACH

When a new technology was applied in an African country to an existing operation, the owner quickly eliminated the employees who had openly not bought into the process. On the surface that seemed like a sensible approach, but among those who remained, some only bought into the technology on the surface and resented it, and others who were fearful of their jobs kept silent. It was a setup that was doomed to failure.

People have their own ways of getting back, like the jungle wife who puts bugs in her dominating husband's soup. In a nutshell, *if your employees don't want something to work, it will not work.*

It is extremely important that you work to provide a team that is sold on the technology. Eliminating naysayers is not good enough. It is better to confront the critics and try to turn them around. The worst thing that can happen is that the technology is questioned to the point that all doubts are eliminated, perhaps bringing the negative players on side in the process.

CONTROL OF THE TECHNOLOGY

It is absolutely necessary for the project leader to have control of the technology and control of its implementation. In the early days of column flotation, for example, the client insisted on changing the control methodology in the test columns, even though the inventors felt certain that the change would not work. Deeper pockets prevailed. The resulting disaster set back the inventors several years.

In another case, the site management and procurement of components was in the hands of the owner, not the seller of technology. The equipment components bought, such as pumps, drives, and air compressors were substandard, severely limiting the success of the project. The quality of site work was limited, because the owner's personnel were doing the work. The ore that had been tested for the design of the plant was from an ore shoot that had petered out and been abandoned and was not the same as the ore delivered to the plant. Changes that needed to be made could not be done, because the owner had diverted the funds to other projects. The process could not be made to work as intended and was prematurely shut down.

BACKUP PLANS

The first SX–EW plant had a backup plan that could be quickly implemented if the process did not work or was deficient in capacity. Right at the outset it is essential that a "What if?" program be developed to encompass all possibilities of success or deficiency.

Probably the easiest way to introduce new technology is to develop a side stream in an existing plant, with plans for abandonment in the case of absolute failure. Vat leaching, for example, as a side process to heap leaching, may prove the case sufficiently to build larger units and abandon the heap.

We mention success in the same sentence, based on experience, where the performance of a process trial greatly succeeded its targets. What happens if you plan for a 1,000 ton per day throughput and you find out the new process is capable of doubling that without adding new equipment? Is there room in the economics for doubling production? Is the plant now constipated at the feed or discharge end, or both? Do you have enough tailings pond capacity for a double-sized plant? Do your feed stockpiles suddenly appear anemic? Do you now have a comminution plant that seems pathetic?

If, on the other hand, your new plant has a 50% shortfall, what are the steps to be taken to improve capacity or add new equipment? Sometimes, it may be necessary to supplement a new process with an old one, just to have a backup stream.

Before a shovel is put in the ground, it is imperative that you gather the team together and run through a "what if?" scenario. This is difficult to do when everyone is on board and pumped up to begin construction of the plant. One last critique, however, is an absolute necessity.

COMPARISON OF DEPLOYMENT VERSUS EMERGING MENTALITY

Like it or not, the days of autocratic decision making are past. Recent years have witnessed the usefulness of emergence mentality (i.e., a team approach to development and application of new technology).

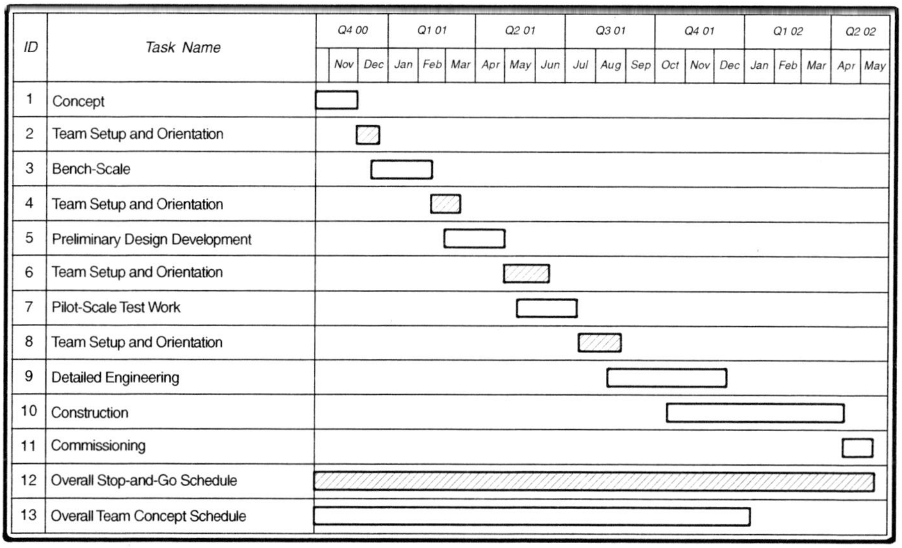

FIGURE 1 Benefit of one project team concept

Autocratic deployment seeks specification of tasks translated into a wide variety of broadly based specifications by individual experts. The goal of each member of the task force is to comply with those specifications. Interfacing requires early fixing. Leader power demonstrates maximum control. Information is channeled and overhead is generally high. Innovations are fewer with generally lower design quality, less enthusiasm, and potentially, several launching problems. The student at the back of the class who may know the answer is seldom asked for an opinion.

In coming years, however, with opportunities to use interfacing hardware and operator "interface" increasing, an emergence mode may become more appropriate. Here the entire design can be let loose to allow all members of the team address constraints. The goal can be to obtain system consistency. Interface will be to mutually maximizing self discipline with information shared under "broadcast" mode. Overhead can be lower, providing many innovative opportunities, with better design quality resulting in more team enthusiasm and fewer launching problems. The team will be a more hands-on energetic team with management providing freedom to experiment with open communication. Efficient technology transfer with the project team–from concept to commissioning–can be beneficially realized under these conditions.

Figure 1 illustrates deficiencies in conventional stop-and-go scenarios. With a fresh team assembled at the end of each activity technology transfer is less efficient, discontinued, has less vision, lack of understanding of end product need and partially lost market and revenue can result. As a case in point, a project carried out by the author lost the market when plant was commissioned with several stop/go situations with changing partners and with a need to reassemble the project team after several time intervals.

At the end, what was demonstrated as a technically viable project, resulting in a product exceeding market specifications, lost the business opportunity. The problem was that the window of opportunity for the project was very narrow, and shortly after plant commissioning, plans for operations had to be terminated because of the changed market conditions.

SUMMARY AND CONCLUSIONS

With the mining industry facing severe pressure from economic and environmental constraints, survival will depend on the application of new technologies. New technology is inevitable; it is time the industry faced up to that and learned to control it. It is also time to break the owner-engineer-financial cycle of restraint that exists.

Control of the technology must be absolute. Too often, the control shifts to the "Golden Rule" (whoever has the gold makes the rules), where the person or group holding the purse strings dictates changes to the technology.

Addressing the people issues is an essential element of introducing new technology, and effective communications is a big part of that. With increasing use of electronic interfacing and the resultant broadcasting of ideas, the team approach becomes much more viable than it has been in more autocratic days.

It is imperative that a backup plan be formulated to deal with surprises—either pleasant or unpleasant—in the application of the new technology.

Finally, a team approach in transfer of new technology is recommended, including those institutions that provide the money and the parties that control the money.

. . . .

Technology Transfer—Perception Versus Reality

Deepak Malhotra*

ABSTRACT

Western technology is considered a panacea for all ills in developing countries. Is that really the case? Are inventors of new technology overselling? What are the expectations of the receiving party? The author offers his views on these questions and proposes a process for systematic transfer of technology while minimizing surprises on the part of the seller and the receiver.

INTRODUCTION

Webster's Dictionary (Tenth Edition) defines technology as "systematic treatment of art" or "improvement of existing system." It can also be considered as a process for handling a specific technical problem. According to these definitions, technology encompasses all methods and processes that can improve the existing process or system.

The objective of implementing new technology is to improve the existing process or build a more efficient new plant. It appears that the grass is always greener on the other side of the fence. Western mining companies believe that developing countries have higher grade resources and an unlimited pool of cheap labor, giving them an unfair advantage in the marketplace. On the other hand, the mining companies in the developing countries think that western countries have an advantage over them because of readily available modern technology.

Is western technology a panacea for all ills in developing countries as they perceive to be true? Are the inventors overselling to the buyers? Other speakers in this symposium are addressing the technical aspects of technology transfer. This paper discusses the sociopolitical aspects of technology transfer and an approach to minimize surprises and enhance the success rate of technology transfer from one sociopolitical setting to another.

EVOLUTIONARY VERSUS REVOLUTIONARY TECHNOLOGY

One can gather from the definition of technology that it is constantly changing. Process engineers are continuously striving to find better technical or economic methods of doing the same job. The new technologies can be either revolutionary or evolutionary.

---

* Resource Development, Inc.

TABLE 1 Examples of evolutionary and revolutionary technologies

Evolutionary	Revolutionary
High rate thickeners	Heap leaching
Large flotation cells	SX/IX*
Pressure filters	Process control
Column flotation cells	Bioleaching
Mobile crushing system	In situ leaching
Flash flotation	Carbon-in-pulp

* SX/IX: solvent extraction/ion exchange

Evolutionary changes result in small improvements to the existing process. These types of changes generally result in replacement of or modification to one unit operation. The risk–reward ratio is low and it is easy to commercialize the technology. On the other hand, revolutionary technologies are based on discarding the existing principles and developing processes based on new principles. This type of technology is significant and results in quantum improvement on the existing process, but because it is unproven, the risk–reward ratio is high. Generally, it is difficult to implement this technology within an existing plant without making major modifications as well as a significant capital investment. Table 1 gives examples of revolutionary and evolutionary technologies in mineral processing.

TECHNOLOGY TRANSFER

The technology transfer can take place from one sector to another within a given nation or from advanced to developing countries. Most of the sociopolitical considerations are the same in both cases. British economist Charles Cooper defined technology transfer as "the transfer or exchange from advanced to developing countries of the elements of technical know-how which are normally required in setting up and operating new production facilities and which are normally in very short supply or totally absent in developing economics" (Cooper 1970).

During the discussions about technology transfer, it is important to take the values of the developing country into consideration. Values refer to "attitudes, preferences, styles of life, normative frame works, symbolic universes, belief systems, and networks of meaning which human beings give to life" (Gould 1989). Instrumentation of the plant may be a desirable goal but may conflict with the values of the host country if providing jobs is an important mission for the company or country, or for both.

EXPECTATIONS OF BUYER AND SELLER

It is important to understand the expectations of the buyer as well as the seller of technology irrespective of the location of the two parties (i.e., developed or developing countries). Table 2 summarizes these expectations. Both parties have expectations that generally conflict with each other.

The buyer becomes aware of the new technology by reading technical publications or through "hearsay." Generally, these papers are written by the inventor of the technology. They present a glowing review of the new method of solving the problem. The buyer convinces his or her management to buy technology by stating that this is the miracle drug that will cure all ills in the plant. Generally, the expectations are unrealistic. The buyer, looking for immediate gratification, assumes that the technology is well developed and flawless and that the supplier is very knowledgeable.

TABLE 2 Expectations of purchaser and supplier of technology

Purchaser	Supplier
Cures all technical problems	Buyer knowledgeable about what seller is pushing
Improvement in plant performance	Short training period (i.e., 2 weeks)
Quick return on investment	Trained personnel involved
Seller knows all	Short construction/start-up period
Seller to train personnel	
Seller is responsible for technology working	
Expect technology to work well	**Sell product and walk away**

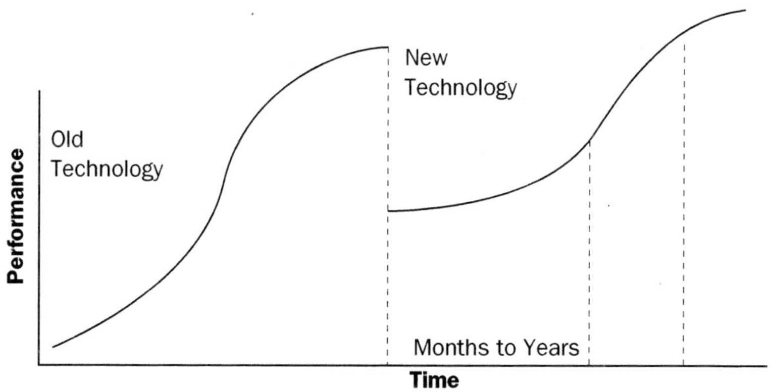

FIGURE 1 Learning curve for new technology

The supplier is motivated by eagerness to sell the technology and assumes that it is a small piece in a big puzzle. The supplier believes that he or she is dealing with knowledgeable buyers who know a lot about the process. Suppliers know that it may take time to optimize the technology but generally do not inform the buyer about that.

It is evident that the expectations and goals of the two parties are significantly different. Both parties generally underestimate the learning time required for the new technology to working efficiently. A significant drop in performance during the learning period for the new technology can result in plant performance lower than that of the old technology for a prolonged period (Figure 1). The buyer becomes impatient and concludes that the new technology is no good. This results in conflict with the seller who had significantly different expectations from what he encountered during technology transfer. Several examples of conflict have been seen when technology transfer has taken place within the western world or within a single country (such as the United States). The problems are compounded when the value systems of the two countries differ significantly.

HOW TO MINIMIZE CONFLICTS?

Many conflicts are traceable to technology itself rather than to the mechanism by which it is transferred from one regional setting to another. The eagerness on the part of the buyer and the seller results in both overlooking the sociopolitical considerations in technology transfer. No matter how careful you are, you cannot put a square peg in a round hole.

The following action plan is recommended for the buyer to minimize conflicts with the seller of the technology:

- Train employees to look at things differently. At the same time, hire a consultant familiar not only with the new technology but also with cultural aspects of the two regions (i.e., West/East).
- Evaluate and optimize the existing process.
- Establish the need for the new technology and determine what the results or consequences of implementing this technology would be.
- Establish realistic performance expectations for the new technology.
- Contact the supplier of the new technology.
- Visit plants that use this technology and discuss problems encountered and performances obtained with operating personnel.
- Develop a realistic learning curve time frame and dedicate appropriate resources for it.
- Review your plans with the technology supplier. Make sure the supplier buys into your plans.
- Have the consultant be the liaison for the technology transfer.

Despite the above recommended action plan, remember that Murphy's law will still apply—"If things can go wrong, they will."

- If the technology transfer takes place from the western world to developing countries, additional factors should be taken into consideration:
- Consultants should be knowledgeable in cultures and languages of the two countries.
- Technology implementation should not result in replacing workers with machinery if providing jobs is a priority in the host country.
- Local trained personnel should not only be able to operate the technology but should also be able to provide plant maintenance.
- A minimum stock of spare parts should be imported if the location is remote and if government red tape interferes with placing orders for the designed equipment.

If these factors are overlooked, the new technology may result in building a "Taj Mahal" that may become a nonoperating monument to remind us that technology can indeed be a double-edged sword, at once beneficent and destructive.

CONCLUSIONS

Successful technology transfer requires understanding sociotechnical as well as sociopolitical aspects of both sides. Technology is not necessarily a panacea for all ills of underdevelopment. It may be beneficial only if it meets the realistic expectation of the buyer. At best, its promise is uncertain.

REFERENCES

Cooper, C. 1970. The transfer of industrial technology to the underdeveloped countries. In *Institute of Development Studies Bulletin*. Brighton, UK: University of Sussex.

Gould, D. 1989. *The Uncertain Promise*. New York: New Horizon Press. 342 pp.

Mentoring: An Important Tool for Engineers and Their Organizations

Leonard Harris* **and Val Ness†**

ABSTRACT

Mentoring is defined as close counseling by an experienced and objective advisor within the industry. It is one aspect of attracting the best candidates into the minerals industry, holding their interest, and encouraging stable career growth.

By encouraging proper mentoring practices, we foster strong careers, we transmit important wisdom to the next generation, we fortify the industry with better talent, and we encourage the best possible advocates for the minerals industry.

This paper illustrates the consequences of poor mentoring, offers an example of positive mentoring practice at the college or university level, reviews some guidelines for stable and fruitful mentoring, and proposes a plan to spread appropriate mentoring practices.

Authors' note: The use of masculine pronouns is for convenience and simplicity only. Our best engineers and managers come from both genders, and the authors intend no slight to either one.

THE NEED FOR MENTORING

Our professions in the mining and mineral industries are in danger of extinction, particularly in North America. This paper is directed at one of the many aspects comprising this problem—obtaining, retaining, and training the best and brightest advocates for the mining industry.

It is a primal need of the human condition to see that knowledge and experience are passed on to the next generation. In traditional cultures, trades are passed from father to son, and from mother to daughter, for generation after generation. In developed cultures, this training is institutionalized through educational systems, apprenticeships, and other formats.

Mentoring is another practice, one that can extend this educational framework throughout the professional career. Mentoring is on point, providing help with the most urgent issues, and it supplies useful feedback for both the protégé and the mentor. And very importantly, it has a powerful appeal for the talent that this industry so desperately needs to fill the ranks in the coming years.

* Mining Industry Consultant

† Jacobs Engineering

COMMON PRACTICE

In common practice without mentoring, the ideal candidate for a junior engineering position with an operating company will have a suitable baccalaureate degree and perhaps some introductory experience with summer internships. This demonstrates that he is trainable, tractable, and qualified as "fresh meat" for the operating environment. However, it is unlikely that his formal training has included any details on issues such as:

- The politics and culture of this company or industry
- Career paths within the organization
- Informal expectations that were not part of the job description
- Organizational history—good, bad, and ugly

At consulting and engineering firms, junior engineers usually have the magic 3 to 10 years of operating experience, supplied, of course, by someone else. Under ideal circumstances, the new hires are brought on board just in time to start work on a fresh project. Nowhere in this career path is there any explanation of:

- The nature of the consulting engineer's work
- Typical project organization and procedures
- Tried-and-true work methods

A similar course of events occurs with manufacturers' representatives and other organizations that supply goods and services to the minerals industry. In any scenario, the unsuspecting engineer is usually thrown into the fray, without a clue, to pick up this new profession as he may. Obviously, this procedure is expedient, but it also has several more subtle advantages.

Self Esteem

Having been brought so suddenly into unfamiliar territory, the new hire remains off guard, allowing more experienced coworkers and supervisors to perpetuate the sensation by feeding the victim contradictory tips on procedures, methods, and company gossip. This helps to maintain the senior employee's position in the organizational food chain, and helps to ensure that the new employee looks like an idiot by comparison.

Self Reliance

Everybody who meets the new employee will realize that this is an awfully confused person, who might not make it through the month. Their reaction will branch in one of two ways: They will treat the "newbie" as though he has a communicable condition, and have nothing to do with him. In that case, he generally displays self reliance, and makes up his own way of doing things.

Adaptability

Alternatively, they will demonstrate great sympathy for the poor wretch, and earn personality points by instructing him in their own idiosyncratic way of running the show.

Camaraderie

As this repeats itself with nearly every new warm body that crosses the portal, everybody learns to enjoy the sideshow. Thus a jolly good time is had by all, more or less like a pack of hyenas taking down a wounded antelope. General morale is thereby improved.

Meanwhile, the new employee is encountering no one with a positive view of his role in the industry. He will not need many weeks or months of this kind of environment to realize that the 15-hour-per-day schedule of the pre-initial public offering (IPO) startup across town looks pretty good by comparison.

But this situation is not limited to that of new employees. For each of us, as we step up the ladder, we encounter new situations where we would benefit from the advice of someone who has been there before. This issue is one that permeates organizations, and one where immense benefits can be gained by appropriate mentoring practices.

It should not be uncommon for a given individual to be on both sides of different mentoring relationships. Taking it a step further, a smart protégé will seek several mentors, individuals who are best suited for advice on separate aspects of his or her developing career. Such a group can act like an ad hoc advisory board, greatly improving the protégé's probability of success.

AN ENCOURAGING EXAMPLE

The following example is adapted from an address to the 1994 metallurgical graduating class at the Lima, Peru, University of Engineering, which covered these points.

Undergraduate Mentoring

During their undergraduate studies, it is important to impart certain information to mining industry student engineers on nontechnical subjects not covered during these studies. This information should cover those items gained by experienced engineers after having worked for several years in the industry.

Self Confidence

At graduation, most of us tend to underestimate ourselves. The students need to be assured that they will be as good as any other professionals at graduation, that they are being well prepared at their school to face the challenges of the industry, and that all they will lack at graduation will be experience.

Speaking Out

They need to be told that they should not be afraid to speak out when necessary, to live up to their principles, and to protect their rights in a firm but respectful manner (and in writing if necessary). They should know that there are people who do not believe in allowing subordinates to speak out and that in those cases, such action may lead to discipline or even termination of employment. It would be prudent, therefore, to keep some funds in the bank to allow for a search for another job if this happens. The alternative is to close their mouths, thereby sacrificing their beliefs and principles.

Flexibility

Three events in life are inevitable: death, taxes, and change. The students should know that change is not always necessarily for the good, but is nevertheless inevitable.

Richard Paulin, former president of the Mineral Economics and Management Society (MEMS) noted in his speech at the 9th Annual Conference of MEMS (Boulder, Colorado, April 12, 2000) that:

> *Change has always been a fact of life. What's new is the rate of change. It's faster than ever before and it's affecting every aspect of life and society: science and technology, organizations, economy, governments, and culture. The mining industry in the year 2000, like most industries on the planet, is evolving at the speed of light.*

No one knows what the changes will be, but it is of prime importance to the students to realize that there will be changes during their lives. This understanding will help them cope with the changes when they do come.

Those of us born 50 to 70 years ago have seen great changes since we graduated as engineers—in the areas of computers, television, jet planes, nuclear energy, microwaves, spaceships, satellites, plastics advanced composite materials, genetically engineered products, the fall of the Berlin wall and communism, the Internet, and the graduation of female engineers. In the mining industry, we have seen the advent of 300-ton trucks; 50-cubic-meter shovels and loaders; remote controlled underground jumbo drills; autogenous and semiautogenous grinding mills; 150-cubic-meter flotation cells; dump and heap leaching; solvent extraction-electrowinning; pressure and bacterial leaching; adsorption of gold and silver on activated carbon; ion-exchange resins for uranium recovery from solutions; and oxygen-assisted roasting and smelting. We have also seen many disasters that were not predicted, and that resulted in deaths and injuries in our industry.

To handle such changes, students will need to be flexible, embracing the changes and not living in the past. They should not tie their careers to any given metal or commodity, but be prepared to switch products, employers, and locations if it is good for them. As engineers, they need to be very aware that they may have to move to inhospitable places throughout the world.

Working Hard

The saying that "hard work never hurt anyone," coined many decades ago, applies equally today. Working hard does not mean, however, that anyone should work 12 hours per day 7 days per week. The idea is to work hard, but work intelligently. For these students, it is imperative to stress the importance of a well- organized workload that allows them to efficiently carry out their jobs, yet leave time for leisure for themselves and their families.

Management Principles

The students need to always keep in mind the basic management principles—planning, organization, control, and delegation. Like any other business, mining is expected to be profitable. To ensure that this happens, and that their work plays an important role in this profit making, continuous short- and long-term planning is required, including:

- Annual and monthly production planning
- Planning to ensure an ample supply of ore for processing
- Maintenance planning to ensure maximum equipment performance
- Cost planning to set budgets and production goals
- Planning for capital expenditures and expansion or improvement projects
- Planning for meetings
- Planning for vacations

To ensure that plans are met, organization and control systems are required. In the mining business, production and cost plans are of vital importance. Budgeting is one of the most important means of organizing and controlling. It is also the easiest way of evaluating a person's performance. Any engineer involved in setting up a budget will be expected to continually monitor and control it for his own good, as well as for the good of the company.

Productivity tables or graphs, along with continuous studies of major cost drivers such as labor, materials, and supplies are also very important measures of performance. Some examples are:

- Tons of ore treated per man-hour
- Ounces, kilograms, or tons of product produced per man-hour
- Power consumption per ton of ore treated
- Reagents consumed per ton of ore treated
- Unit costs of fuel, explosives, tires, cyanide, lime, and grinding balls, among others

Delegation is the fourth management principle. The students should know that no one can handle everything on and off the job by himself. In the words of the former chairman of the board of Newmont Mining Corporation, Plato Malazemoff, "no one has a monopoly on brains." This becomes very important when one is promoted to supervisor or superintendent or manager. One needs to surround himself with good people–share the load and delegate.

The students should know that even with all this planning, organization, control, and delegation, it will be necessary on occasion to work extra hours without being paid for overtime. Our profession is not one of 8 hours per day 5 days per week. There will always be reports to be written, plant emergencies to address, and problems to be solved. This sometimes requires work on birthdays, holidays, anniversaries, or when others are partying.

The students should also be told that the search for knowledge does not end on graduation. It will be necessary to continue studying by taking courses, attending seminars and conferences, reading technical bulletins, and visiting mining operations.

The Positive Attitude

One of the easiest things to do in life is to take the negative attitude. With this attitude, there is no need to do anything. On the other hand, if the positive attitude is taken, some action and some work will be needed. The students should be told that during their careers, they will hear such things as:

- "We tried that, it doesn't work."
- "It's too expensive."
- "It takes too long."
- "We have no experience in that field."
- "Our people won't accept that."
- "That's not the way we do it."
- "You are new here, and you know nothing about the way we do things around here."
- "What do you know about metallurgy. You are a geologist, aren't you?"
- "What do you know about mining. You are a mining engineer, aren't you?"
- "We can't do it."

Solving problems and making progress requires that a positive attitude be adopted. If it is important, it *has* to be done. It will be up to *all* concerned to find the way to do it, as in the saying, "Where there is a will, there is a way."

When there is a problem, it is important not to give up. The students should know that they should seek help from their books, from their peers, from their superiors, and from their subordinates. It is mandatory to involve subordinates by always talking to them, encouraging them to speak up and give feedback. They should never be muzzled but treated in a fashion like the engineer would like to be treated by his superiors. All involved should work together as a team.

Protecting the Environment

The students should know that the mining industry does not have a good reputation. Those who are unfamiliar with the industry often view miners as a bunch of polluters and believe that mining is dirty, degrading to the environment, and dangerous. Things that were done in the past, known by some today as SOOP (sins of our past), should not be repeated in the future. The students must be conscious of the past errors in mining, and be sure that the future will be different. They will ensure that mining can be conducted anywhere, while being at the forefront of environmental protection.

Safety

Likewise, mining does not have a good reputation for protecting its employees from accidents on the job. The students should be instructed to always protect themselves and all others from harm on the job, and they must deliberately involve themselves in safety courses and join safety teams. Furthermore, safety has a real effect on the bottom line. Responsible safety practices will lead to stronger balance sheets.

Social Responsibilities

The students should know that an important part of doing business today in the mining industry, especially overseas, involves community assistance programs for the people who live around the mining areas. The purpose is to build up mutual cooperation and respect so that the company and its neighbors may live and work in peace. The students should be aware of this important need and should set aside some effort to help their fellow humans both on and off the job.

Sense of Humor

Finally, the students should keep up their sense of humor and have fun. It is not uncommon to hear from some engineers who have left the mining industry to remark that "it is not fun anymore."

By paying attention to and abiding by their principles and the technical aspects of their profession, and by the acceptance of the "transferable or soft skills" (by Dr. Paulin) noted above, the students may look forward to successful and happy professional lives.

MENTORING CHECKLIST

The relationship between the protégé and his mentor may be short term or long term, but there should be a set of governing ground rules. First, the two parties should agree on the frequency and location of the meetings, so that they are mutually convenient and neither party carries an undue burden. Second, each party will have a set of expectations for the other to fulfill.

Mutual Points of Agreement

- Frequency and location of meetings
- Broad goals of the relationship
- General structure of specific goals and assignments
- Approximate time to be committed by both parties
- Metrics for evaluating the relationship
- Expected duration of the relationship

Protégé's Expectations of the Mentor

- Mentor's willingness to listen as well as advise
- Mentor's help and guidance in finding good projects
- Regular work assignments from the mentor
- Mentor's help with getting constructive feedback from others
- Mentor's help in figuring out the organizational culture, personalities, and politics
- Mentor's helpful tips about how things work and unwritten rules
- Mentor's candor and constructive feedback at all times

- Mentor's insights into a viable "roadmap" for professional development and progress in the organization
- Mentor's help in "tooting the horn" of the employee at times of notable success
- Tips on how the mentor would have handled a project if it were his
- Strict confidentiality

Mentor's Expectations of the Protégé

- Punctuality at each meeting
- Candor, honesty, and open dialogue
- Forethought about help the mentor can provide
- Willingness to listen and take constructive criticism
- Willingness to act on advice once it is accepted
- Willingness to take personal responsibility for his own career

Finally, if it isn't working, either party must say so!

PLAN FOR IMPLEMENTATION

As a mentor, you can expect to gain a high degree of satisfaction in helping someone develop, and you will enjoy the exposure to new ideas. But from the perspective of the protégé, it may be difficult to find an appropriate mentor for a given issue. As senior members of your professional community, if you are interested in taking on this role, you will have to signal your availability. Speak to the junior members in your organization, especially those not directly in your chain of command. But be sensitive to political issues, to avoid offending those who are in direct supervisory roles. Alternatively, it can be especially rewarding to deal with individuals outside of the organization, thereby avoiding most of the political issues, and potentially opening broader avenues for the exchange of ideas.

Some companies are building mentoring programs into their human resource departments, or they are making mentoring a formal responsibility of their managers. Many of these programs have established baseline requirements for mentors and protégés, including training sessions where participants discuss expectations and generate procedures for fulfilling them.

Another established mechanism is through the SME Student Mentoring Program, which is coordinated through Barb Filas and Joette Cross of Knight Piésold. By contacting them, you can be placed on a list of available mentors for undergraduate and postgraduate students who are anticipating their careers in mining and earth sciences.

Again, mentoring can address a small but vital part of the current dilemma in the minerals industry. To strengthen this industry, we will need good talent, and to obtain and keep the good talent, widespread and effective mentoring practices are a part of the solution. But it is far from a panacea, and our expectations should be tempered accordingly on both sides of the table.

BIBLIOGRAPHY

Dixon, Susani N.H., Esq., Holland & Hart LLP, Denver, Colorado, 2000.

Hart, Lois, Women's Leadership Institute, Lafayette, Colorado 80026, February 2000.

Digges, Diana, "Not Your Fairy Godmother," <http://www.careerbuilder.com/achieve/ach_0007_godmother.html>

Intercultural Assessment, Training, and Development: A Must for International Assignees and Their Families ©

Michael F. Tucker, Ph.D., CMC

ABSTRACT

Dealing successfully with intercultural issues is critical for success on an international assignment. This is true not only for employees but for the accompanying spouses and children as well. This paper presents a set of research-based competencies required for successful life and work in another country, such as open-mindedness, flexibility, and respect for the beliefs of others. Best practices for training employees and their families bound for international assignments are also presented.

INTRODUCTION

John Smith and his family had just arrived in Indonesia, where he was assigned to an important management position by his U.S.-based mining company. The mine in Indonesia had started up successfully during the past year, but there continued to be management problems between the Americans and Australians on the one hand and the Indonesian employees on the other. Smith had been told that it was very difficult to trust Indonesians, because "they only tell you what they think you want to hear, and they are very reluctant to bring you bad news quickly." Sure enough, an incident occurred that led Smith to believe this was true. Despite initial meetings with members of his Indonesian staff, where he made it very clear that he expected to "be told the truth quickly," he just wasn't getting the information he needed. One day there was an accident that caused injury and significant expense. Smith's inquiry revealed that members of his Indonesian staff had known about the problem for some time but had not told him.

Ralph Adams was working at his U.S. company's mine site in Peru. Adams' Peruvian counterpart, Ignacio Dominguez, had been in the United States before Adams was assigned to Peru. Adams had not taken much time to spend with Dominguez in either country because of what he perceived to be schedule and task priorities. Dominguez had invited him to dinner, but Adams had just not been able to find the time.

---

* President, Tucker International

The project that Adams and Dominguez were working on involved selecting a Peruvian contractor for a critical project. Adams wanted to use the U.S. company's procedure for competitive selection of contractors. Dominguez had a very different approach. He said, "I intend to use a contractor that I have worked with for a long time. I have a good relationship with the head of the company, who is my cousin, and I know that he will do a good job." Adams explained how important it was to get the best contractor at the lowest price. "The U.S. company's approach will allow us to look at several suppliers and pick the one that best meets our needs," he said. Dominguez said he already knew the other Peruvian suppliers and assured Adams that his cousin's company was the best.

After several more meetings, it was clear that Dominguez was simply not going to use the competitive bid process. Adams felt very uneasy about Dominguez and about his cousin's company.

These two cases illustrate the critical importance of understanding and managing the cultural differences involved in an international assignment. In the Indonesian case, Smith did not deal successfully with certain aspects of the Indonesian culture that involved power and authority, high versus low context of communication, and direct versus indirect messages.

Power and Authority

Indonesians do not like to bring bad news to someone in authority. For one thing, it implies that the boss is at fault because he or she is supposed to be on top of everything. Indonesians view those in authority quite differently than do Americans, accepting a greater degree of inequality between those in charge and the less powerful. Dutch researcher Geert Hofstede (1991) has studied this aspect of cultural differences and has created a dimension called "power distance." On a scale of 11 (low power distance) to 104 (high) Indonesians score a 78; Americans score a 40. The Indonesians with higher power distance have a society with an autocratic style and a tight hierarchical structure where individuals know their place and the limit of their roles. They were not about to break out of this limited role and bring Smith the bad news directly.

High- Versus Low-Context Communication/Direct Versus Indirect Messages

There was a way for Smith to gain access to the information he needed in a timely way, but he should not have expected direct communication. Indonesians have a high-context culture where *how* something is said, to whom it is said, and in what circumstances it is said is as important as the message itself. Americans are low context, where value is placed on direct communication focusing on the content, not the context, of the message. The Indonesians believed that they would have "lost face" and damaged their relationship with Smith if they had brought him this bad news so early on in their relationship. Smith might have been more successful had he established a "third party" in the communication network–someone both he and the Indonesians trusted who could deliver these kinds of messages with little damage done.

In the Peruvian case, Adams was not aware of the "polychronic" nature of the Peruvian culture as compared with the "monochronic" American culture. This aspect of culture was identified and studied by Edward T. Hall (1990). He noted that polychronic cultures are relationship-oriented, and have a great involvement with and value placed on time spent with people. Monochronic cultures, on the other hand, are more objective. They see time as something to manage efficiently and do not like to be distracted by other people. The polychronic Peruvians value close, long-term relationships and often operate quite naturally within a wide network of extended family and friends. What Adams viewed as inappropriate and possibly a case of nepotism was a natural way of doing business for Dominguez. In fact, Dominguez's cousin may well have been the very best choice for the contract work because his performance would reflect the trust placed in him by Dominquez. It is likely that he would do his very best to honor the trust placed in him so as not to lose face or to cause

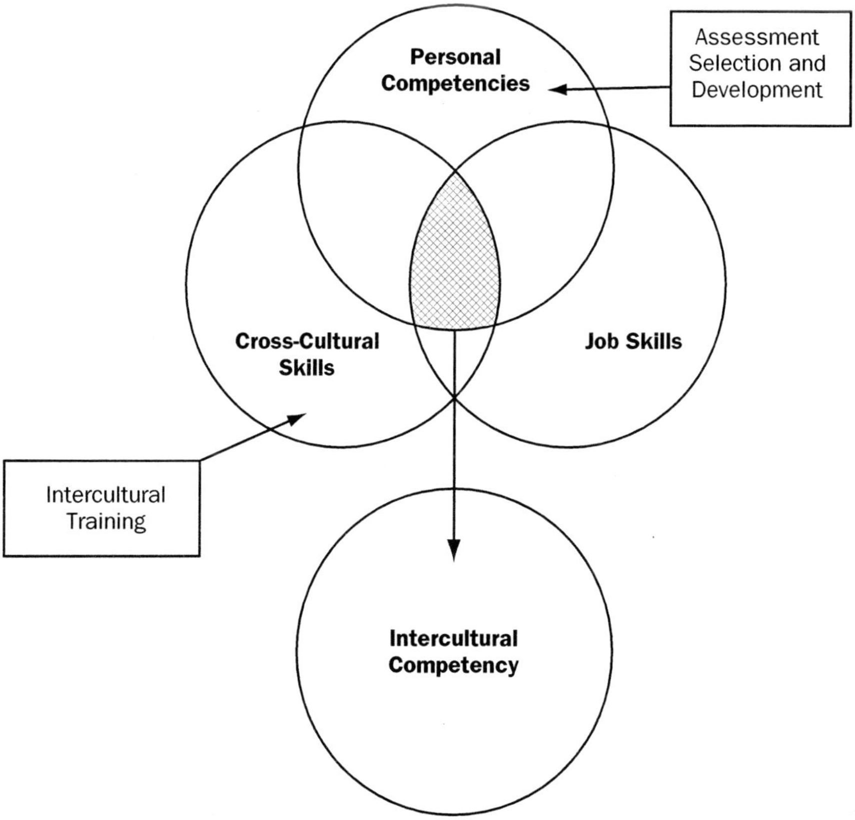

FIGURE 1 Intercultural competency model

Dominguez to do so. Adams could have managed this situation much better had he known this critical difference between American and Peruvian cultures. He could have spent social time with Dominquez and built a relationship with him. He could have also spent time with Dominguez's cousin and then judged for himself what capabilities his company represented, instead of viewing the situation from his own cultural perspective.

INTERCULTURAL COMPETENCY

By developing intercultural competency through a professional process of assessment, training, and development in preparation for their assignments, international assignees and their families can manage the issues illustrated in these cases more effectively. It is simply not enough to have the requisite technical, managerial, or administrative skills to succeed on an international assignment. Intercultural competency is also required, as shown in Figure 1.

This model of intercultural competency is applicable to employees, although a similar model is applicable to spouses and accompanying young people. Families who are sent on international assignment should be treated as a unit, and the special needs of family members should be incorporated into the professional preparation process. It is well known that an expatriate spouse can make or break an international assignment. If the spouse becomes well-adjusted and happy, the employee can apply full attention to the job. If not, the entire assignment may be disrupted. Involving the spouse and the family in the assessment and training process can go a long way toward ensuring success.

Intercultural competency also has a great deal to do with successful transfer of technology across cultures. Many expatriate assignments require this transfer to host country colleagues. Research (Hawes and Kealey 1979) and practice indicate that successful intercultural teaching, coaching, and mentoring requires deep adaptation to the local culture and very good intercultural communication. Many countries that are recipients of technology transfer have relationship-oriented cultures, and their people cannot learn from even the best experts if they cannot accept them and relate to them as people.

ASSESSMENT AND DEVELOPMENT FOR SUCCESS ON INTERNATIONAL ASSIGNMENT

The first step in developing intercultural competency for an international assignment is to ensure that candidates for international assignment carefully consider their situation to determine if they are mobile and ready. A useful tool for this type of assessment is the IMA, or International Mobility Assessment (www.tuckerintl.com).

This assessment helps candidates to:

- Honestly assess their current situation to determine if they are ready to take on the challenge of adapting successfully to a different culture
- Make sure that their expectations are realistic about an international assignment, a new job, and especially a new environment and culture

Interactive exercises guide the user through self assessments of international readiness in three critical areas:

1. The family
2. A new country and culture
3. A different workplace

This assessment concludes with a decision sheet, in which all the reasons for pursuing an international assignment are weighed against all those for not going forward. Whatever decision is made, candidates and their families are assured that they have participated together in a high-quality process that can so dramatically affect their lives.

A second type of assessment and development that leads to intercultural competency is to ensure that assignees have the personal competencies required to adapt to life and work in a different culture. This assessment can best be done by means of a well-researched and validated instrument like the OAI, or Overseas Assignment Inventory (www.tuckerintl.com). The OAI has been the subject of continuous research for 27 years, involving more than 10,000 people. It has proven to be a reliable and valid predictor of intercultural adjustment and expatriate success. The areas of assessment in the OAI are illustrated in Figure 2 and then described in the sections that follow.

MOTIVATIONS

The most successful expatriates are those who take on an international assignment with motivations that are:

- Positive or forward looking rather than escaping from something
- Well balanced between job or career and personal development
- Sustaining in ways that can withstand the pressures of life and work overseas

Examples of positive, well-balanced, and sustaining motivations are:

- "I have long wanted to live and work in (the assignment country). I am interested in the people and culture, and want to see how well I can do there."

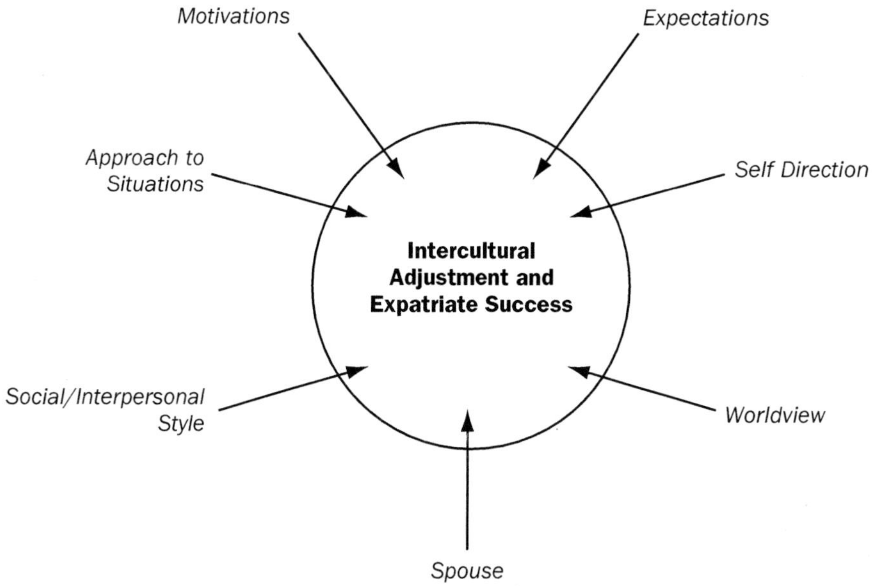

FIGURE 2 Personal intercultural competencies

- "I think that this assignment is a great opportunity for my family to experience a broader lifestyle and to grow and learn together."
- "This international assignment offers me a great career opportunity and the job is an important one for the company. But it is also a professional and personal development opportunity."

On the other hand, poor motivations are those that focus primarily on the money to be made, or on getting away from a negative work or personal situation.

EXPECTATIONS

Those who are realistic about what it will be like to live and work in another country, its associated difficulties, as well as the probable benefits, have a greater chance of success than those who have low expectations and do not look forward to the opportunity. If one expects to succeed and looks forward to the assignment and country, while aware of the challenges, one is likely to adjust and adapt favorably to the new surroundings.

SOCIAL INTERPERSONAL STYLE

Trust in People

Meaningful personal and professional relationships in another country will develop if expatriates can convey and encourage mutual trust among coworkers and business associates.

Interpersonal Interest

Experience has shown repeatedly that the human element of interpersonal relationships–in this case, intercultural people skills–is critical to an expatriate's success and happiness overseas. Those who are sincerely interested in, accepting of, and concerned for others have a great advantage in adjusting to another culture.

Social Adaptability

This predictor pinpoints the ability to socialize comfortably with new people in new and unfamiliar social situations and to accept and be accepted by new groups of friends and acquaintances. A loner or someone who feels comfortable only in a small, intimate group may feel lost in a new and unfamiliar setting.

WORLDVIEW

Open-Mindedness

Open-minded individuals are receptive to different beliefs and ideas without feeling as if their own are being challenged or threatened. Those with the attitude that their own or their country's way is inherently superior will face difficulties in accomplishing many tasks.

Respect for Other Beliefs

The capacity to be nonjudgmental of other people's religious and political beliefs is extremely important in another cultural environment. Expatriates who demonstrate a willingness and ability to respect and be interested in the beliefs of other cultures are more likely to establish meaningful intercultural relationships.

Tolerance

Effectively adapting to another cultural environment requires an ability to interact with, or live closely to, people who may have fundamentally different habits and lifestyles from one's own. This also means withstanding living conditions and surroundings that are different from or less comfortable than one is accustomed.

APPROACH TO SITUATIONS

Flexibility

The ability to consider new ideas and to realize that there is more than one valid way to approach and solve a problem is necessary for effective intercultural adjustment.

Patience

Expatriates need to understand that a "sense of time" means different things in different cultures, or else they may be paralyzed by frustration from unexpected delays. Expatriates must remain patient when business protocol demands a seemingly roundabout decision-making process or way of doing business.

Sense of Humor

A good sense of humor is one of the most overlooked, yet important, aspects of effective intercultural coping and adjustment. The ability to bring humor into difficult or confusing situations, and to laugh at and learn from one's own mistakes often helps to ease tensions and facilitates communication.

Initiative

Those who succeed on international assignments are self-starters, meaning that they do not sit back and expect things to happen or rely on someone else to take care of matters.

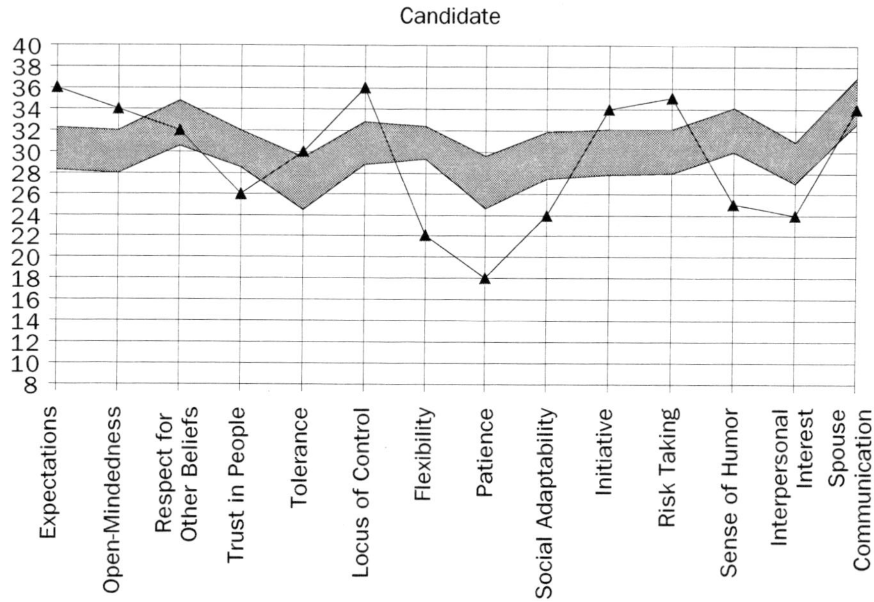

FIGURE 3 Example of an overseas assignment inventory profile

Risk Taking

Life in a new country requires exploring new things and learning new ways of doing things. The willingness to take risks, meet challenges, and cope with change greatly enhances intercultural adjustment.

SELF DIRECTION

Locus of Control

Expatriates who believe that they can control, shape, or direct the course of their lives are likely to exert more effort to make things work abroad. Those who believe that things happen because of luck or fate often will feel helpless when confronted by new and changing life situations.

COMMUNICATION BETWEEN SPOUSES

When communication between couples is open and constructive, relationships are often enhanced and strengthened by the overseas experience. At the same time, however, the unique stresses associated with adjusting to a new culture can be very difficult and damaging to weak or troubled relationships.

An example of an OAI profile for a candidate being considered for an assignment in Indonesia appears in Figure 3. The blue band is the current database. Any score in this band is average, above is a strength, and below is an area of concern. Although strong in several OAI dimensions, this particular candidate would represent considerable risk for assignment to Indonesia because of scores below the norm in trust in people, flexibility, patience, social adaptability, sense of humor, and interpersonal interest. These areas are important competencies needed to adapt to the polychronic culture of Indonesia, where a soft approach to relationship building is necessary.

INTERCULTURAL TRAINING FOR DEVELOPMENT OF CROSS-CULTURAL SKILLS

The third and final circle in the intercultural competency model is the development of cross-cultural skills through intercultural training. This type of training has become an accepted part of the preparation of expatriates. Surveys by the Employee Relocation Council (ERC) and the Society for Human Resource Management (SHRM) show that the majority of international organizations now provide intercultural training. There is a great variance, however, in the quality of such training. Best practices indicate that this training should be at least 3 full days in length, research-based, staff-intensive, participative, and well evaluated.

Research-Based Training

In addition to researching the needs of participants to ensure best fit and training relevance, training programs should be based on field research that is proven to lead to successful intercultural adjustment. A six-factor model based on field research is presented below. These six factors are the goals and expected outcomes of intercultural training (www.tuckerintl.com).

Acceptance. Those who accept the culture of the country of assignment show respect for local customs and behavior patterns. They do not criticize or make light of the culture, but accept it as different from their own but entirely natural for local people.

Knowledge. Successful expatriates are genuinely interested in their country of assignment. They learn historical and contemporary information about the country and are able to engage in conversation with local people about subjects that are of interest to them.

Affect. Successful intercultural adjustment leads to feelings of well being. These feelings in turn are associated with a positive self concept and positive attitudes about the country and its people.

Lifestyle. Expatriates who adjust well lead a very active and rewarding lifestyle. They are able to do some of the things that they enjoyed back home as well as engage in activities that are unique to their country of assignment.

Interaction. Successful adjusters engage themselves in the country of assignment, which means that they choose to be with local nationals not only on the job, but during their discretionary time as well. They make local friendships that replace those left back home and that help support their new lifestyle.

Communication. Intercultural adjustment is closely associated with intercultural communication. This means learning the language as well as time, business, and other constraints allow. Learning the nonverbal communication system of the local culture and using that system to demonstrate respect, acceptance, and understanding is also important.

Staff-Intensive Training

Many intercultural training programs are conducted on a private basis, for one expatriating family at a time. Even these small programs should involve the following staff:

- Highly qualified master trainer
- Area studies expert on the country of assignment
- Host-country business culture consultant
- Belief systems consultant
- Host-country resource people
- Returned expatriate spouse
- Assessment and development consultant
- Highly qualified youth trainer

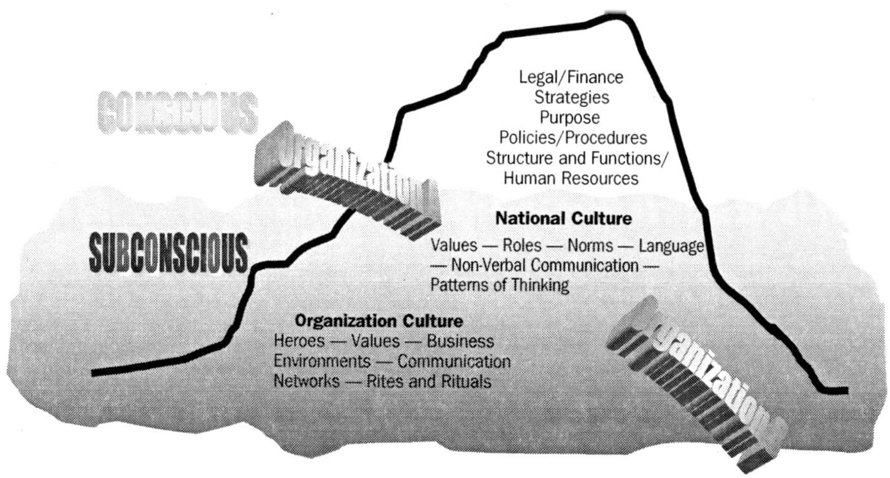

FIGURE 4 The Iceberg Model of Culture

Participative and Instrumented Training

Training should be based on solid adult-learning principles of participation and instrumentation.

- Content-rich lectures should allow for comment, questions, and reflection.
- Discussions should focus on participant interests and needs and not on those of the training staff.
- Case studies, critical incidents, cross-cultural analysis exercises, and simulations should be used.

Sample Training Modules

Culture as a Framework for Learning. The "Iceberg Model of Culture" appears in Figure 4. This analogy of an iceberg (adapted from Berger and Luckmann 1966) is a useful way to distinguish between "surface culture" and "deep culture" and to begin exploring the deep national and organizational cultures of the assignment country. As illustrated in the figure, only about 11% of the mass of an iceberg appears above the water; some 89% of it remains hidden below the surface. When applied to culture, things that are observable or that occur on the conscious level are referred to as surface culture. What is not observable, or subconscious, is called deep culture.

Intercultural Values Exercise. A central part of intercultural training is direct discussion with host-country nationals. Values are discussed and contrasted between home and host cultures (see Table 1).

Business Culture in the Country of Assignment. The topics presented in Table 2 are included in a presentation and discussion by an expert in the business culture in the country of assignment.

Focus on the Expatriate Spouse. A discussion with a spouse who has successfully managed a "portable life" for the family in the country of assignment covers the following topics:

- Social conventions
- Education
- Working
- The law

TABLE 1 Intercultural values exercise

Personal Control Over the Environment
Change
Time and Its Control
Equality/Egalitarianism
Individualism and Privacy
Self Help Concept
Competition and Free Enterprise
Future Orientation
Action/Work Orientation
Informality
Directness, Openness, and Honesty
Practicality and Efficiency
Materialism/Acquisitiveness

TABLE 2 Business culture

Business Themes and Their Cultural Underpinnings
Business Implications: Management Practices/Styles and Business Communication
—Task/Goal Orientation
—Decision Making and Problem Solving
—Business Meetings
—Business Communication
—Persuasion, Conflict Resolution, Criticism, Accountability, and Blame
—Leadership and Management
—Initiative, Risk Taking, Dealing with Change, and Asking for Help
—Ambition, Motivation, and Work Ethic
—Performance Reviews
—Status and Rewards
—Business Ethics
Workplace Environment and Socializing
—Women and Minorities
—Attitudes Toward Foreigners
—Business Socializing and Entertaining
—Corporate Citizenship

- Entertaining
- Clothing
- Child rearing and education
- Health care

Intercultural Communication. Effective intercultural communication is a core requirement for successful international assignment. A module on this subject includes the following:

- Communicating in a high-context culture
- Names and titles
- Direct versus indirect messages
- Communicating bad news with "soft landings"
- Nonverbal signals
- Importance and use of a third party

Intercultural Training for Youth. Providing intercultural training programs for expatriate children and young people is highly important and necessary. Families are units. By reducing fear and stress and creating a more realistic and optimistic view about the international move, young people are happier and more supportive, adjust easier, and demonstrate a willingness to culturally engage themselves in the country of assignment. It is necessary to counsel and educate children on how to appropriately handle changes that greatly affect their lives.

The following are the top ten concerns of young people who move to another country:

- Leaving friends
- Making new friends
- Leaving school behind
- What to expect about the new school and fitting in
- Leaving pets behind
- What life will be like
- Not understanding the language
- Being removed from sports activities
- Not liking the new country, climate, customs, and food
- Being afraid

Intercultural training for youth helps them to manage these concerns. It covers fundamental issues surrounding culture, culture shock, moving, and transition. Through discussions and exercises, the program addresses their fears, hesitations, excitements, and worries. They also learn a great deal about their new country in an interactive way.

Training Program Evaluations

Because of the importance of intercultural training in terms of time and resource allocation, these programs should be well evaluated. Immediate post-program evaluations by participants yield their reactions to the program and their assessment of learning achieved. In addition, follow-up evaluation in the host country should be done 9 to 12 months after arrival to result in:

- Direct evaluation of training from the perspective of life and work in the country of assignment
- Evaluation of intercultural adjustment
- Feedback of results to revise training design

Benefits of Intercultural Training

People who participate in effective intercultural training report that the program:

- Teaches them a great deal of essential information in a very short time
- Reduces anxiety about relocating and adjusting to a new job and home
- Offers insights into their own capabilities and areas necessary for expatriate success that are in need of development
- Provides a framework for continued learning during the expatriate assignment
- Facilitates a more seamless and shorter period of adjustment to the country of assignment

International organizations that supply this training have realized benefits including:

- A higher assurance that their investment in expatriate assignments is not lost through early return or poor performance

- A shorter time for expatriates to become fully capable in the assignment, as well as enhanced performance during the assignment
- A higher level of acceptance of expatriates among offshore affiliates
- An excellent cost-benefit ratio, when the program fee is compared to the total cost of an expatriate assignment
- Corporate representatives who are more competitive in key global locations

REFERENCES

Berger, P.L., and T. Luckman. 1966. The Social Construction of Reality: A Treatise in the Sociology of Knowledge. New York: Anchor Books-Doubleday.

Hall, E.T., and M.R. Hall. 1990. *Understanding Cultural Differences*. Yarmouth, Maine: Intercultural Press.

Hawes, F., and D. Kealey. September 1979. *Canadians in Development: An Empirical Study of Adaptation and Effectiveness on Overseas Assignment.* Ottawa, Canada: Canadian International Development Agency, Communications Branch Briefing Centre.

Hofstede, G. 1991. *Cultures and Organizations: Software of the Mind.* London: McGraw-Hill.

How Specific-Equipment Training for Maintenance Personnel Pays Dividends for New Facilities

Clifford O. Hamilton*

ABSTRACT

In both mature and young maintenance workforces, the workforce commonly lacks some of the skills required for their assigned work. We call this the "skills gap." This skills gap usually exists in a new operation because although you may hire qualified maintenance personnel, they more often than not have little if any experience with the specific equipment at the facility. Performance Associates International, Inc., has developed and proven a technique to address the maintenance skills gap that is usually found in a new operation.

INTRODUCTION

During the last 20 years or so, most new mining facilities have recognized the need to positively address the operational training requirements associated with new facilities. Today's metallurgical process plants depend greatly on the "knowledgeable worker" who has a conceptual understanding of the overall process. The advent of distributed control systems (DCS), combined with the ever-increasing dependence on sophisticated processing of lower grade materials, has forced the mining industry to approach the training of operating personnel differently than has historically been the case. Until 20 or so years ago, the general approach to process operator training was for a more experienced older operator to take the new operator around and show him the "ropes."

This approach to operator training has been replaced with a much more formalized, targeted training approach that emphasizes the basic engineering and scientific principles that underlie the overall process control scheme associated with today's metallurgical process plants. The mining industry has accepted, and more often than not positively addresses, the need to train plant operators so that they conceptually understand what is happening in the process. Implementing this approach to training has resulted in most new plants realizing tight control over process variables, with the resultant reduction in the number and duration

* Vice President, Performance Associates International, Inc.

of process upsets. This has also resulted in the production of quality products at acceptable operating costs.

However, the approach to training the maintenance workforce for today's process plants is still generally grounded in the "old methods." It is usually assumed that the maintenance personnel will be hired from a pool of experienced maintenance people who have a good grounding in fundamental maintenance skills. As a result of this assumption, very little effort is directed at ensuring that maintenance personnel are properly prepared to maintain the highly sophisticated mechanical, electrical, and instrumentation equipment found in the process facilities of today. Generally, maintenance training for new facilities is limited to what some equipment suppliers offer as part of their after-sales effort. Additionally, all too often, new facilities incorporate the newest maintenance techniques, which include condition monitoring, operations and maintenance integration, and total productive maintenance, among others. These are all good and valid concepts. However, unless the maintenance workforce is equipped with the required baseline of knowledge about the equipment to be maintained, maintenance effectiveness will be seriously compromised.

IDENTIFICATION OF MAINTENANCE TRAINING REQUIREMENTS

In our experience, a maintenance workforce can often lack some of the skills that their assigned work requires, which is called the "skills gap." This gap usually exists in a new operation. Even though you may hire qualified maintenance personnel, they may have little if any experience with the specific equipment that was purchased for the facility.

A new metallurgical plant organization must recognize that it may face such a skills gap in its maintenance workforce. It is highly unlikely that each maintenance employee, whether supervisory or hourly, possesses all the required basic maintenance skills and specific-equipment skills for each piece of equipment that is used at the particular facility.

We believe that the best approach to training is to define the job tasks required of employees in each classification and area. The skills required of each employee directly follow the tasks. For example, if a mechanic must remove and replace a Warman pump, he or she must be able to align the coupling while installing the new pump. Additionally, the specific type of Warman pump that is in place at the new facility may be a new model that requires specific preventive maintenance that is totally unknown to the maintenance personnel hired for the facility. In the aforementioned example, a potential skills gap exists in terms of basic skills and specific-equipment skills.

Unlike operating personnel, maintenance personnel require a number of technical skills that are not intuitive to personnel who have not had technical training. Examples of these skills include the ability to use certain tools, a knowledge of electricity, the ability to weld, and a knowledge of lubricants, among others. These skills are generic in nature, and they are commonly referred to as "basic skills." The basic skills capabilities of the maintenance workforce are normally assessed with some form of testing. Based on an applicant's previous work experience and how he or she scores on the basic skills tests, a reasonably informed assessment can be made. However, specific-equipment training must almost always be developed and implemented anew at each new metallurgical facility.

SPECIFIC-EQUIPMENT TRAINING PROGRAM DESIGN

Specific-equipment training means instruction on the specific equipment or instrumentation, or both, installed in the plant. This training assumes that maintenance personnel have a command of the required basic skills of their craft. Because this training is very specific to the plant being constructed, this training must be custom-developed, utilizing vendor documentation. Specific-equipment training also must include training on the preventive maintenance (PM) procedures for the equipment. These PM procedures must be developed for the specific equipment in the plant. Additionally, training on the use of various parts and service manuals should be incorporated into the specific-equipment training.

The training program for each job classification should be built in a series of modules, each providing training for a group of associated skills. Figure 1 is an example of a training module data sheet for a process plant mechanical maintenance supervisor's recommended specific-equipment training.

In Figure 1, the recommended training program content for the process plant mechanical maintenance supervisor includes training in the following areas:

- Crushing, conveying, and stacking specific-equipment training
  - Fixed crusher
  - Mobile crusher
  - Rock breaker
  - Apron feeder
  - Conveyors
  - Agglomerator
  - Stacker
- PM procedures
  - Service intervals
  - Tasks performed

Although the typical mechanical maintenance supervisor of a process plant has the basic skills necessary to oversee the maintenance of the process plant equipment, there is usually a need to upgrade his or her specific-equipment maintenance knowledge.

Figure 1 details the process plant mechanical maintenance supervisor training program, which includes specific-equipment training on the crushing, conveying, and stacking equipment. This training includes:

- A detailed description of the unit and how it is to be operated
- Proper PM and corrective maintenance procedures for the equipment
- Details on how to properly start up and shut down the unit
- Techniques for troubleshooting the unit when breakdowns occur

The training program reference material for the specific-equipment training is normally extracted from vendor manuals. In addition, custom PM checklists are developed using the vendor manuals.

Once this training is completed, the supervisor will be well positioned to manage and ensure that proper maintenance is performed on all crushing, conveying, and stacking equipment.

Figure 2 illustrates the recommended training program content for a mechanical maintenance hourly worker in a process plant, based on the same plant equipment as shown in the example training module data sheet for the mechanical maintenance supervisor.

Similar to the recommended specific-equipment training for the mechanical maintenance supervisor, the specific-equipment training for the mechanical maintenance hourly personnel, shown in Figure 2, includes a detailed description of the unit and how it is to be operated. The training also includes proper PM and corrective maintenance procedures for the equipment, as well as details on how to properly start up and shut down the unit. Finally, it covers the techniques that should be used to effectively troubleshoot the unit when it breaks down.

In addition to the formal or classroom instruction, we normally recommend a period of on-the-job training (OJT) for all hourly maintenance personnel. The recommended OJT component consists of performing actual work on the job, using the skills associated with

Training Block:	Supervisor's Plant Mechanical Maintenance Training (SPMM)	Training Duration (Hours)			Reference Material Develop or Procure		Training Module Development		
Module Code	Module Name & Subheadings	Formal	OJT	Training Reference Material	Hours	Resource	Hours	Resource	Instruction Resource
SPMM001	*Crushing, Conveying/Stacking* *Specific-Equipment Training* Fixed crusher Mobile crusher Rock breaker Apron feeder Conveyors Agglomerator Stacker	32		Extractions from vendors' manuals	140	Performance Assoc.	24	Performance Associates	Client organization trainer
SPMM002	*PM Procedures* Service intervals Tasks performed	8		Custom PM checklists	80	Performance Assoc.	24	Performance Associates	Client organization trainer
Totals		40	0		220		48		

FIGURE 1 Example training module data sheet for a mechanical maintenance supervisor

Training Block:	**Process Plant Mechanical Equipment Training (PPME)**	Training Duration (Hours)			Reference Material Develop or Procure		Training Module Development		
Module Code	Module Name & Subheadings	Formal	OJT	Training Reference Material	Hours	Resource	Hours	Resource	Instruction Resource
PPME001	*Crushing, Conveying/Stacking Specific-Equipment Training* Fixed crusher Agglomerator Mobile crusher Rock breaker Apron feeder Conveyors Stacker	16	80	Extractions from vendors' manuals	40	Performance Associates	24	Performance Associates	Client organization trainer
PPME002	*Crushing, Conveying/Stacking Mechanical PM* Review of checklists Field identification of requirements	8	32	Custom PM checklists	80	Performance Associates	24	Performance Associates	Client organization trainer
Totals		24	112		120		48		

Callout (arrow pointing to OJT value 80): Recommended on-the-job (OJT) training duration in hours

FIGURE 2 Example training module data sheet for mechanical maintenance hourly personnel

TABLE 1 Typical training module

Module Section and Explanation	Comments and Examples
1. Module Objectives and Outline	
—List of each specific objective to be accomplished with the training module	When the training is complete, the trainee should be able to accomplish each objective
—Outline of what is to be covered in the training module	Outline includes suggested time intervals and reference designations
2. Reference Material	
—Information to be learned and used as a reference for training	The vendor documentation and PM checklists provide information on:
The reference material for specific-equipment training is taken from the vendor documentation and custom PM checklists. It also includes an overhead transparency of any relevant graphics contained in the reference material	—The description of the unit —Details how it is to be operated —Proper PM procedures —Proper corrective maintenance procedures —Startup and shutdown —Troubleshooting
3. Workbook	
—Questions the student must answer while referring to the reference material (see above) —An answer sheet is provided for the instructor	Questions are generally fill-in-the-blank or, in some cases, short answer paragraphs. The workbook is designed to reinforce lectures. It is a training device, not a validation device
4. Knowledge Assessment Test	
—Questions the student must answer without reference to any material, i.e., a closed-book test —An answer sheet is provided for the instructor	The test validates that the trainee has mastered the subject theory. Multiple choice and true/false questions are used
5. Qualification Checklist	
—Checklist to be completed by the trainee's superior, validating that the student has satisfactorily achieved the necessary practical skill(s)	The qualification checklist validates that the trainee can satisfactorily apply the theory on the job

each of the applicable modules. Normally, the OJT component is defined at the time each module is developed. It consists of a listing of the tasks and the total hours that must be completed on the job for each applicable module. The OJT component serves to reinforce the formal training, along with validating that the student can actually apply the theory in the field.

Once the recommended specific-equipment training program content has been defined for each job classification, maintenance specialists develop training modules for each equipment item at the new facility. Table 1 shows the format of each of the maintenance training modules.

The module normally consists of an outline of its contents; a series of objectives to be met; and the materials to be used in conducting the training, such as a book, overhead transparencies, or other visual aids; workbooks to be completed by the trainees; and normally, a test to validate that the module was successfully learned. In addition, a qualification checklist is included, which is completed by the trainee's immediate supervisor once the trainee returns to the job. The qualification checklist verifies that the maintenance trainee can directly apply the theory associated with each major learning objective during his or her work on the job.

Once the modular structure is defined and the modules are developed, a determination is made for each employee as to exactly which modules are required. Using a test in conjunction with the employee's previous work experience is usually the recommended methodology for determining which modules an employee will need to take.

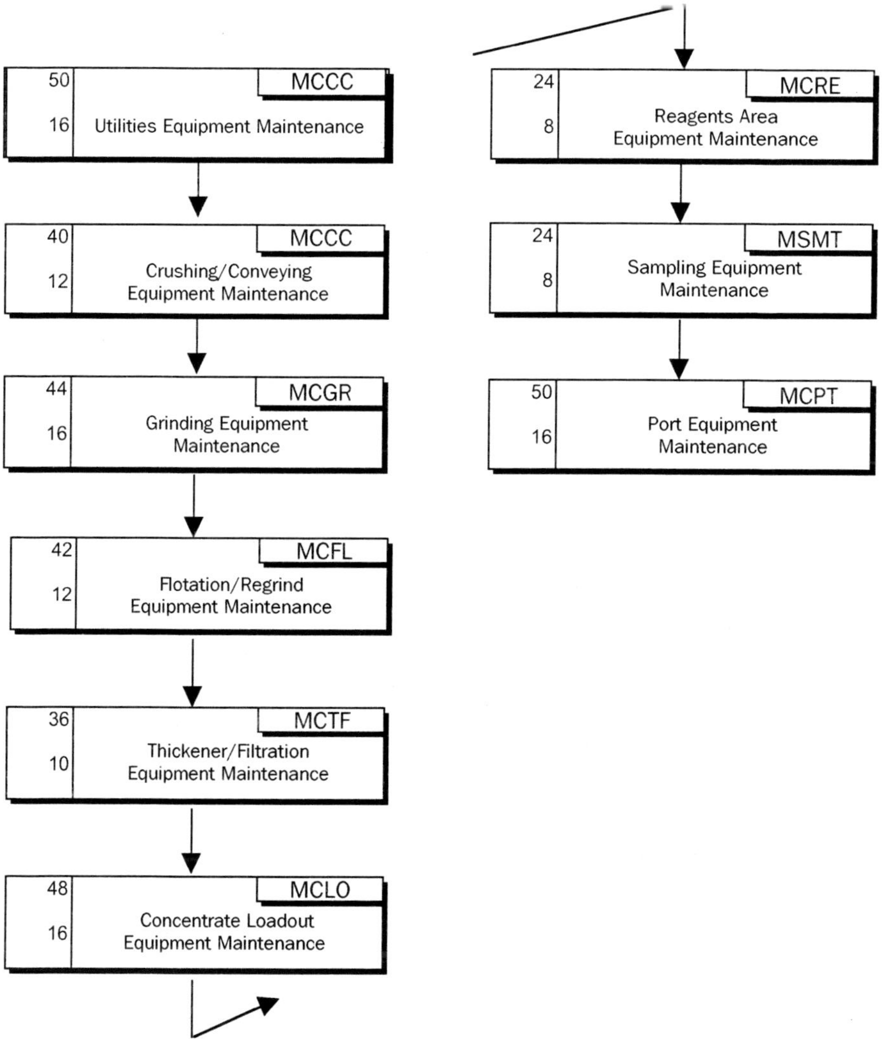

FIGURE 3 Plant mechanical maintenance hourly personnel learning path

The modular training structure for each area job classification is organized in a path such that each employee can progress from one module to another. This type of training organization allows for the fact that the maintenance employee may not need every module—greatly reducing the possibility of training personnel in subjects that they already know.

Figure 3 illustrates an example of a typical learning path for a mechanical maintenance hourly worker. The symbology used in the learning path diagram in Figure 3 is explained in the learning-path block-symbol legend shown in Figure 4.

The hours in the upper left corner of Figure 4 indicate the formal classroom instruction hours for the block. Note that formal instruction may include field training under the direct supervision of an instructor. The hours in the lower left corner of the block diagram indicate OJT. The employee is not fully qualified until he or she has satisfactorily completed the OJT hours and the immediate supervision has signed off on a qualification checklist.

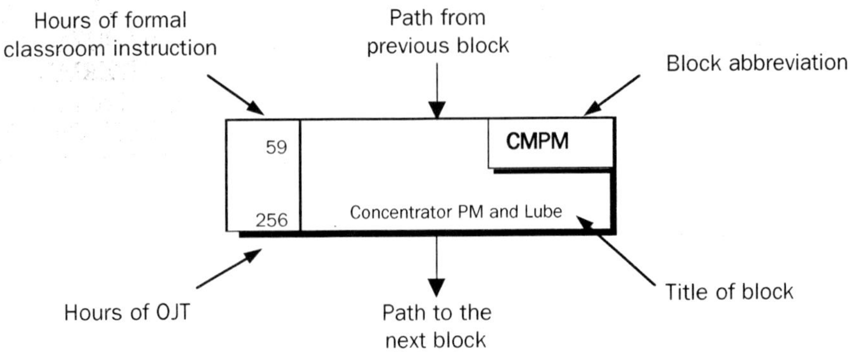

FIGURE 4 Learning-path block-symbol legend

THE DIVIDEND FOR PROPER MAINTENANCE TRAINING IMPLEMENTATION

Today's modern metallurgical facilities rely on high equipment availability to achieve profitable results. Equipment downtime associated with poorly maintained facilities, or facilities that contain equipment with which maintenance personnel are unfamiliar, or both, has to be eliminated if anticipated returns on investment are to be realized.

Allowing maintenance personnel to learn how to properly maintain and troubleshoot equipment in the field without the benefit of a well-structured specific-equipment training program is not an effective way to proceed.

In our experience, the technique described in this presentation of developing specific-equipment training modules, combined with classroom instruction, has assisted operations in eliminating the trial and error methods employed when this type of training approach is not taken.

Yes, the project incurs costs for this type of training. But as a wise man once said, "If you think education is expensive, try ignorance."